T0291855

Cambridge Tracts in Mathematics and Mathematical Physics

No. 58

STOCHASTIC APPROXIMATION

To
My Father and Mother

STOCHASTIC APPROXIMATION

M. T. WASAN

Professor of Mathematics
Queen's University, Kingston, Ontario

CAMBRIDGE
AT THE UNIVERSITY PRESS
1969

PUBLISHED BY THE PRESS SYNDICATE OF THE UNIVERSITY OF CAMBRIDGE
The Pitt Building, Trumpington Street, Cambridge, United Kingdom

CAMBRIDGE UNIVERSITY PRESS
The Edinburgh Building, Cambridge CB2 2RU, UK
40 West 20th Street, New York NY 10011–4211, USA
477 Williamstown Road, Port Melbourne, VIC 3207, Australia
Ruiz de Alarcón 13, 28014 Madrid, Spain
Dock House, The Waterfront, Cape Town 8001, South Africa

http://www.cambridge.org

First published 1969
First paperback edition 2004

A catalogue record for this book is available from the British Library

Library of Congress Catalogue Card Number: 69–11150

ISBN 0 521 07368 5 hardback
ISBN 0 521 60485 0 paperback

CONTENTS

PREFACE

The subject of stochastic approximation is of recent origin, but the number of research papers which have appeared in theoretical and applied journals over the last fifteen years, speaks for its practical utility and theoretical importance. I had the following motives in preparing this manuscript.

First, I have collected a number of applications of stochastic approximation to problems in engineering and medicine and have shown how one can make use of these techniques in practice. Thus the manuscript can have practical values.

Secondly, since the material of the manuscript is not available in text-book form except for a chapter in a sequential methodology book in statistics or mathematics, this can be used for one semester course for graduate students in mathematics, computer sciences and mathematical statistics.

Finally, this can serve as a reference text for research workers in science, engineering, mathematics, statistics and many other fields.

I cannot precisely state the prerequisites for studying this subject, but a reader should know the subject matter of the appendices. Usually a proof of a result is carried out extensively, but in order to avoid monotony, in a few places proofs are kept concise where they are already fairly long. Appendices 2 and 3 are of great help in these cases.

I am grateful to Professor D. L. Burkholder who introduced me to the field of stochastic approximation by his interesting course conducted in the Fall semester of 1956 at University of Illinois. Since then I have kept my interest alive by application of these techniques to industrial problems, doing research and conducting a course at Queen's University.

It is a great pleasure to extend my thanks to Professor R. A. Bradley who encouraged me to convert my lecture notes into a text during his visit to Queen's University under the IMS visiting lecturer programme. I also thank Professors R. H. Farrell, M. M. Rao and many others for their helpful suggestions and comments in preparation of the manuscript. I am grateful to Professor V. Fabian for reading the manuscript and making several valuable suggestions, and I extend my appreciation to Professor J. F. C. Kingman who edited this manuscript with patience and great care, which improved the manuscript, particularly, its presentation.

I am grateful to the Queen's Art Research Committee for the grant and the University of Bombay for the facilities. Finally I express my thanks to Mr H. R. Chavan and Mr M. B. Joshi for typing the first draft of such a difficult manuscript, and to Mrs E. M. Wight for the final draft.

M. T. W.

January 1968

CHAPTER 1

INTRODUCTION

1. Introduction

Because of the fast development in computer technology and its applications, there is a great deal of interest in making use of applied mathematics and in developing new techniques to meet new situations. Numerical Analysis is such an area, particularly iteration techniques. These are old and well-known mathematical tools with widespread applications in both theoretical and applied work. However their mathematical counterparts, stochastic approximations, are of relatively recent origin and in a state of rapid growth.

Professor H. Hotelling discussed many ideas of stochastic approximations in his paper of 1941 [46] and relevant results have been given by Friedman and Savage [38] and others. But Robbins and Monro [63] in their pioneer paper gave a formal mathematical treatment of this field and proved many interesting results. Since then many papers on the subject have appeared in theoretical and applied journals, indicating its usefulness and growing importance.

In this chapter in order to motivate the subject, a number of practical problems are stated as illustrative examples. Some interesting examples are also given in the chapter of applications. A method of stochastic approximation is defined, and is then compared with another sequential method the so-called up-and-down method, and with iterative techniques of Numerical Analysis. Some of its merits and demerits which indicate situations where stochastic approximation can be exploited with advantage are appended. In the last section of this chapter we summarize the main results of each chapter.

2. Illustrative examples

(*a*) It is known that hardness of copper–iron alloy is influenced by the length of time the alloy is aged at 500 °C. Let x be this length of time and $Y(x)$ be the hardness of the alloy, the problem is to find the values of x which can give an alloy of a given average hardness α. It is a well-known fact that hardness varies from alloy to alloy.

(*b*) Let us consider the sensitivity of explosive to shock. A common method is to drop a known weight of some explosive mixture from a given height, some will explode, others will not. Each specimen has a critical height, the problem is to determine this height.

(*c*) In testing insecticides, one may be confronted with the problem of determining the critical dose for a given quantitative response.

(*d*) Let us consider a plot where x lbs of fertilizer is applied, and $Y(x)$ lbs of corn is produced. The yield will probably be small if a smaller amount of fertilizer is used, but also if too much is used. Somewhere in between the maximum yield will be achieved. Yield will, of course, vary from year to year even though x remains the same.

3. Stochastic approximations

The situations discussed in (*a*), (*b*) and (*c*) can be formalized mathematically in the following manner. In the region of experimentation an experimenter chooses arbitrarily a value x_1, conducts an experiment and observes the value $y(x_1)$ of the random variable $Y(x_1)$ with expectation $M(x) = E\{Y(x_1)\}$, where E denotes mathematical expectation and M is some increasing function of unknown form. He chooses also a sequence of positive numbers a_n which decreases with n. For example, he can choose $a_n = c/n$, where c is any positive real number. His problem is to determine the value of θ such that $M(\theta) = \alpha$. He sets up the following recursive relation in order to pick a value of x for his next experiment.

$$x_{n+1} = x_n - \frac{c}{n}[y(x_n) - \alpha]. \tag{1}$$

Suppose he has already conducted the nth experiment and that as a result he knows x_n and the value of $y(x_n)$ Then using (1) he can determine what value of x he should use in the $(n+1)$th experiment. Let us examine this recursive relation. For simplicity let $\alpha = 0$. Then (1) reduces to the form

$$x_{n+1} = x_n - \frac{c}{n} y(x_n). \tag{2}$$

If the value of $y(x_n) > 0$ then $x_{n+1} < x_n$ and if $y(x_n) < 0$ then $x_{n+1} > x_n$. This seems to be reasonable because one is interested in solving $M(\theta) = 0$. If $y(x_n)$ is positive then one should decrease the value of x for the $(n+1)$th stage of experimentation, and vice versa. In Chapter 2 conditions under which the sequence $\{x_n\}$ converges in mean square and with probability one to the solution θ are stated. For the situation (*d*) refer to Chapter 3 for a method of locating the maximum of the regression function $M(x)$.

The up-and-down method, another sequential method, is a relevant competitor to a stochastic approximation method.

4. Up-and-down method

Consider such situations as (*a*), (*b*) and (*c*). Suppose now an experimenter chooses x_1 arbitrarily and also a constant d which is approximately equal to the standard deviation of the random variable. Again he is interested in solving the equation $M(x) = \alpha$. He tests a single subject at any given time. If at the nth step in the process the stimulus is at level x_n, the sequential rule is that the $(n+1)$th test be made at level

$$X_{n+1} = \begin{cases} X_n + d & \text{if there is no response at } X_n, \\ X_n - d & \text{if there is response at } X_n, \end{cases} \tag{1}$$

when the experiment is terminated at some chosen value n. An essential difference between the up-and-down method and the stochastic approximation method is that for the up-and-down method the value of x is changed by a fixed amount d in the direction dictated by the experiment. In some experimental situation it is feasible only to change the value of x by a fixed amount d. An approach to the problem is that of Probit Analysis. A choice of a method to be employed is determined by practical limitations

5. Newton–Raphson method

We now define the Newton–Raphson method, an iterative technique of Numerical Analysis, and discuss its comparison with a stochastic approximation method.

Let M be a function from the real interval $I = [a, b]$ to I, whose form is unknown. The problem is to solve the equation $M(x) = \alpha$. One can do this by the following iteration techniques. Choose x_1 arbitrarily in the interval I and use the following recursive scheme to generate a sequence which will converge to the desired solution.

$$x_{n+1} = x_n - [M'(x_n)]^{-1} [M(x_n) - \alpha], \tag{1}$$

where $M'(x_n)$ is the derivative of M at $x = x_n$. Appendix 1 gives the appropriate conditions under which a solution exists. (1) can be written as follows:

$$x_{n+1} = x_n - a_n[M(x_n) - \alpha], \tag{2}$$

where $a_n = [M'(x_n)]^{-1}$ which is assumed to exist and to be bounded. From equation (1) of §1.3,

$$\left.\begin{aligned} X_{n+1} &= X_n - a_n[Y(X_n) - \alpha], \\ X_{n+1} &= X_n - a_n[M(X_n) - \alpha] - a_n[Y(X_n) - M(X_n)]. \end{aligned}\right\} \tag{3}$$

Equation (3) looks like (2) but with the additional element $-a_n[Y(x_n) - M(x_n)]$ (which may be described as 'noise' arising from the fact that $M(x_n)$ is not exactly observable). Thus in order to have a solution we will have to assume all the conditions of Appendix 1. That is, M will have to be monotone or continuous on the interval $[a, b]$, and satisfy the Lipschitz condition there. In addition we will have to assume conditions under which the 'noise' $-a[Y(x_n) - M(x_n)]$ should disappear as the number of iterations increase. That is essentially what we shall do in subsequent chapters.

6. Applications

We have seen in §1.2 that there are a number of applied problems where stochastic approximation could be employed. But these problems can be solved by other statistical methods, for example, in some cases the method of least squares may be used.

Kushner [55] has concluded that the least-squares method is superior to the stochastic-approximation method after discussing the efficiency of the two methods. The main advantage with the stochastic-approximation method is that one need not know about the input of the system, all one needs to know is the output which is easily available in practice. Furthermore, it is unnecessary to know the form of the regression function or to estimate unknown parameters. Thus stochastic approximation is a non-parametric technique which quite often generates a non-Markov stochastic process.

There are three main problems associated with stochastic-approximation procedure. First, one will be interested in the convergence and mode of convergence of the sequence generated by the method to the desired solution of the equation. Secondly, one would like to know the asymptotic distribution of the sequence. Finally, since stochastic approximation is a sequential procedure, one will be interested to know an optimum stopping rule for a given situation. The first problem is tackled in Chapters 2, 3, 5, and 7, the second in Chapter 6. It appears that these two problems have been answered satisfactorily, but the problem of optimum sequential stopping rules has hardly been attempted. A paper pertaining to the subject is that of Farrell [35].

7. Summary

In Chapter 2 the Robbins–Monro method is described for a particular applied problem of response-no-response analysis. This has motivated the subject of stochastic approximation and reflected what is involved mathematically. Then a general one-dimensional stochastic approximation is investigated. It is shown that when one has some knowledge of the process one can relax conditions. For example, if it is known only that the stochastic-approximation sequence is regular, the conditions imposed on the iterating coefficients can be relaxed. A small sample theory is discussed and comparison is made with the asymptotic one. Then many modifications of a general case are mentioned which have revealing effect on the stochastic-approximation procedures.

Chapter 3 discusses a method of Kiefer and Wolfowitz for determining the location of a maximum of a regression function, and singles out the optimum choice of iterating coefficients. A unified treatment of the subject of Chapters 2 and 3 is given and a procedure for obtaining the location of an inflection point is stated. For a simple regression model, a stochastic-approximation technique is exploited with advantage.

Chapter 4 shows how a method of stochastic approximation can be used in a control problem and its difference from the conventional procedure. Then a problem of pharmocology is discussed. Its application to a reliability situation indicates how such a method is effective in measuring information. Its uses in problems of bioassay are investigated.

In Chapter 5 a multivariate stochastic-approximation procedure is introduced and its utility in the investigation of a kinetic model of pharmocology is discussed.

Chapter 6 deals with the problem of asymptotic distribution and exploits a method of moments and a characteristic function method. An application to the construction of confidence interval is indicated.

In Chapter 7 the techniques of stochastic approximation for continuous random processes are dealt with and their utility for analogue computers is stated. Their applications to a control problem, and a filter problem are discussed.

In Chapter 8 an 'up-and-down' method is investigated and its application to 'Rankits' is shown. Small sample and non-parametric up-and-down methods are appended.

Appendices on the subjects of iterative techniques, limit theorems and inequalities are also added. These are mathematical tools which are exploited in order to develop the theory of stochastic approximation.

The problems at the end of each chapter provide additional information which is obtainable from the references.

REMARK. It will be clear from the context of the material whether we are dealing with a random variable or just its numerical value.

CHAPTER 2

THE ROBBINS–MONRO METHOD

1. Introduction

In this chapter the mathematical aspects of the Robbins–Monro method are dealt with. First, response–no-response analysis is considered; this throws light on the mathematical subtleties involved. Secondly, a general result due to Dvoretzky is proved; this has many modifications, which reveal many ways to economize the iterative procedures.

If one has a regular process, then one can relax the conditions on the iterative coefficients. This is discussed in § 2.4. In practice one generally deals with small sample theory and it is interesting to see its relation to large sample theory, which is discussed in § 2.5.

2. Response–no-response analysis

In this section a simple case of the Robbins–Monro method is considered. This illustrates the mathematical techniques involved in proving a general result and prepares the reader for what he may expect in this chapter.

In many problems of bioassay and applied statistics one obtains response–no-response data from an experiment. For example in in testing insecticides one observes whether or not an insect responds to a certain dose. Thus the problem is to determine the critical dose for a given quantitative response. Mathematically the problem can be formulated in the following way.

Let Z be a random variable with distribution function M. If x is a real number and $Y(x)$ a random variable such that

$$\begin{aligned} Y(x) &= 1 \quad \text{if } Z \leqslant x \\ &= 0 \quad \text{if } Z > x \end{aligned}$$

then
$$\begin{aligned} P[Y(x) = 1] &= P[Z \leqslant x] = M(x), \\ P[Y(x) = 0] &= P[Z > x] = 1 - M(x), \\ E[Y(x)] &= 1 . M(x) + 0 . (1 - M(x)) = M(x). \end{aligned}$$

Now $Y(x)$ is the observable response to a dose x. The problem is to determine the value of x for a given quantitative response α. This can be done as described in the following theorem.

THEOREM 1. *Let M be a distribution function and α a real number such that there is a real number θ giving $M(\theta) = \alpha$; let M be differentiable at θ and have $M'(\theta) > 0$. Let x_1 be a real number and n be a positive integer. Let*

$$X_{n+1} = X_n - \frac{1}{n}(Y_n - \alpha), \tag{1}$$

where Y_n is a random variable such that

$$P[Y_n = 1 | X_1 X_2, \ldots, X_n, Y_1, \ldots, Y_{n-1}] = M(X_n),$$
$$P[Y_n = 0 | X_1 X_2, \ldots, X_n, Y_1, \ldots, Y_{n-1}] = 1 - M(X_n).$$

Then $\lim_{n\to\infty} E(X_n - \theta)^2 = 0$, so that random variable sequence $\{X_n\}$ converges to θ in mean square and hence in probability.

Proof. Let $\xi_n = E(X_n - \theta)^2$ we want to show that $\lim_{n\to\infty} \xi_n = 0$. From (1) we have

$$X_{n+1} - \theta = X_n - \theta - \frac{1}{n}(Y_n - \alpha).$$

Therefore

$$E(X_{n+1} - \theta)^2 = E(X_n - \theta)^2 - \frac{2}{n}E[(X_n - \theta)(Y_n - \alpha)] + \frac{1}{n^2}E(Y_n - \alpha)^2.$$

Let
$$d_n = E[(X_n - \theta)(Y_n - \alpha)],$$
$$e_n = E(Y_n - \alpha)^2,$$

then
$$\xi_{n+1} = \xi_n - \frac{2}{n}d_n + \frac{e_n}{n^2},$$

$$\sum_{j=1}^{n} (\xi_{j+1} - \xi_j) = -2\sum_1^n \frac{d_j}{j} + \sum_1^n \frac{e_j}{j^2},$$

$$\xi_{n+1} - \xi_1 = -2\sum_1^n \frac{d_j}{j} + \sum_1^n \frac{e_j}{j^2}.$$

Since
$$0 \leqslant e_n = E(Y_n - \alpha)^2 \leqslant 1, \sum_1^n \frac{e_j}{j^2}$$

is non-decreasing and bounded, and

$$\sum_{j=1}^{\infty} \frac{e_j}{j^2}$$

converges, because $\sum_1^{\infty} \frac{1}{j^2} = \frac{\pi^2}{6}$,

$$\begin{aligned} d_n &= E[(X_n - \theta)(Y_n - \alpha)] \\ &= E\{E[(X_n - \theta)(y_n - \alpha) | X_1, \ldots, X_n, Y_1, \ldots, Y_{n-1}]\} \\ &= E[(X_n - \theta)(M(X_n) - \alpha)] \geqslant 0 \end{aligned}$$

since $(X - \theta)(M(X) - \alpha) \geqslant 0$ for all X.

Hence $2 \sum_1^n \frac{d_j}{j} = \xi_1 - \xi_{n+1} + \sum_1^n \frac{e_j}{j} \leqslant \xi_1 + \sum_{j=1}^{\infty} \frac{e_j}{j^2}$

and thus

$$\sum_1^n \frac{d_j}{j}$$

is bounded non-decreasing, which implies that

$$\sum_1^{\infty} \frac{d_j}{j}$$

converges. Therefore

$$\lim_{n \to \infty} \xi_{n+1} = \xi_1 - 2 \sum_1^{\infty} \frac{d_j}{j} + \sum_1^{\infty} \frac{e_j}{j^2} = \xi$$

exists. We want to show that $\xi = 0$.

Suppose there is a real number sequence $\{k_n\}$ such that

$$\left.\begin{array}{ll} (1) & k_n \geqslant 0 \\ (2) & d_n \geqslant k_n \xi_n \end{array}\right\} \text{ for all } n,$$

$$(3) \quad \sum_1^{\infty} \frac{k_n}{n} = \infty \text{ diverges,}$$

then $\xi = 0$. To see this, note that (1) and (2) imply that

$$0 \leqslant \sum_1^{\infty} \frac{1}{n} k_n \xi_n \leqslant \sum_1^{\infty} \frac{d_n}{n} < \infty.$$

Now suppose $\xi > 0$. Since $\lim_{n\to\infty} \xi_n = \xi$ there exists a natural number $N > 0$ such that if $n > N$ then $\xi_n > \xi/2$. Thus

$$\infty > \sum_{n=N+1}^{\infty} \frac{k_n \xi_n}{n} \geqslant \frac{\xi}{2} \sum_{n=N+1}^{\infty} \frac{k_n}{n},$$

implying

$$\sum_{n=N+1}^{\infty} \frac{k_n}{n}$$

is finite, a contradiction of (3), therefore $\xi = 0$. It remains to be shown that there exists a real number sequence $\{k_n\}$ satisfying (1), (2) and (3).

Let
$$A_1 = |x_1 - \theta| \quad \text{for} \quad n = 1,$$

$$A_n = |x_1 - \theta| + \sum_{j=1}^{n-1} \frac{1}{j} \quad \text{if} \quad n > 1.$$

Since
$$(X_{j+1} - X_j) = -\frac{1}{j}(Y_j - \alpha)$$

$$X_n - x_1 = -\sum_{1}^{n-1} \frac{Y_j - \alpha}{j},$$

and

$$|X_n - \theta| = \left| x_1 - \theta - \sum_{j=1}^{n-1} \frac{Y_j - \alpha}{j} \right|$$

$$\leqslant |x_1 - \theta| + \sum_{j=1}^{n-1} \frac{1}{j} |Y_j - \alpha| \leqslant |x_1 - \theta| + \sum_{1}^{n-1} \frac{1}{j},$$

since
$$|Y_j - \alpha| \leqslant 1 \quad \text{for all } j.$$

Thus
$$|X_n - \theta| \leqslant A_n, \quad n \geqslant 1.$$

Let
$$k_n = \inf \frac{M(x) - \alpha}{x - \theta}.$$

Clearly $\{k_n\}$ satisfies (1) and $\{k_n\}$ also satisfies (2), since

$$d_n = E[(X_n - \theta)(M(X_n) - \alpha)] = E\left[(X_n - \theta)^2 \frac{M(X_n) - \alpha}{X_n - \theta}\right]$$

$$\geqslant E(X_n - \theta)^2 \inf_{|X-\theta| < A_n} \frac{M(X) - \alpha}{X - \theta}$$

since $|X_n - \theta| \leqslant A_n$,
$$\geqslant \xi_n k_n.$$

Since $$\lim_{x\to\theta}\frac{M(x)-\alpha}{x-\theta} = \lim_{x\to\theta}\frac{M(x)-M(\theta)}{x-\theta} = M'(\theta) > 0$$

there is a $\delta > 0$ such that if $|x-\theta| < \delta$ then

$$\frac{M(x)-\alpha}{x-\theta} > \tfrac{1}{2}M'(\theta).$$

Therefore $$\inf_{|x-\theta|<\delta}\frac{M(x)-\alpha}{x-\theta} \geqslant \tfrac{1}{2}M'(\theta).$$

Suppose N is a positive natural number such that $n > N$ then $A_n > \delta$. Consider $n > N$. If $\delta \leqslant |x-\theta| \leqslant A_n$ then

$$\frac{1}{|x-\theta|} \geqslant \frac{1}{A_n} \quad \text{and} \quad |M(x)-\alpha| \geqslant k,$$

where $k = \min\{M(\theta+\delta)-\alpha\} > 0$. Therefore

$$\inf_{\delta<|x-\theta|\leqslant A_n}\frac{M(x)-\alpha}{x-\theta} \geqslant \frac{k}{A_n}$$

and

$$\begin{aligned} k_n = \inf_{|x-\theta|\leqslant A_n}\frac{M(x)-\alpha}{x-\theta} &\geqslant \min\left\{\inf_{|x-\theta|<\delta}\frac{M(x)-\alpha}{x-\theta},\right. \\ &\qquad\qquad \left.\inf_{\delta<|x-0|\leqslant A_n}\frac{M(x)-\alpha}{x-\theta}\right\} \\ &\geqslant \min\left\{\frac{M'(\theta)}{2}, \frac{k}{A_n}\right\} \\ &\geqslant \frac{k}{A_n} \quad \text{for} \quad n > N_1, \text{ say.} \end{aligned}$$

It follows that

$$\sum_{n=N_1+1}^{\infty}\frac{k_n}{n} \geqslant k\sum_{n=N_1+1}^{\infty}\frac{1}{nA_n} \geqslant \frac{k}{k_1}\sum_{n=N_1+1}^{\infty}\frac{1}{n\log n} = \infty,$$

since $$\log n = \sum_{j=1}^{n-1}\int_j^{j+1}\frac{1}{t}dt \geqslant \sum_{j=1}^{n-1}\frac{1}{j+1} = \sum_2^n\frac{1}{j},$$

$$\begin{aligned} A_n &= |x_1-\theta| + \sum_{j=1}^{n-1}\frac{1}{j} \leqslant (|x_1-\theta|+1)\sum_{j=2}^{n}\frac{1}{2} + \sum_{j=2}^{n}\frac{1}{j} \\ &= \{|x_1-\theta|+2\}\sum_{j=1}^{n}\frac{1}{j} \leqslant \{|x_1-\theta|+2\}\log n = k_1\log n. \end{aligned}$$

Thus (3) is proved, and the proof of the theorem is complete.

Thus one can obtain the solution to this response–no-response problem by the stochastic approximation method.

3. One-dimensional stochastic approximation

Now we consider the question of a general situation where Y is not restricted to the value 1 or 0 but can take any value. A general situation is considered by Robbins and Monro [63] where a convergence in mean square and probability of an iterated value is demonstrated, and Blum [4] has proved under simple conditions the almost sure convergence of the procedure. Kallianpur [48] has also considered this problem. Dvoretzky [29] proves a very general result which covers all the results previously proved, and his proof simplified by Wolfowitz [83] is reproduced here. Derman and Sacks [21] gave another simplification of this result. (See problem 9 of this chapter.)

THEOREM 2. (Dvoretzky). *Let* $\{\alpha_n\}$, $\{\beta_n\}$ *and* $\{\gamma_n\}$ *be non-negative real number sequences such that*

$$\lim_{n\to\infty} \alpha_n = 0, \tag{1}$$

$$\sum_1^\infty \beta_n < \infty, \tag{2}$$

$$\sum_1^\infty \gamma_n = \infty. \tag{3}$$

Let θ be a real number and T_n be measurable transformations satisfying

$$|T_n(X_1, \ldots, X_n) - \theta| \leqslant \max[\alpha_n, (1+\beta_n)|X_n - \theta| - \gamma_n] \tag{4}$$

for all $X_1, \ldots, X_n$. Let X_1 and Y_n $(n = 1, 2, \ldots)$ be real random variables and define

$$X_{n+1} = T_n(X_1, \ldots, X_n) + Y_n(X_1, \ldots, X_n) \tag{5}$$

for $n \geqslant 1$.

Then the conditions $E\{X_1^2\} < \infty$,

$$\sum_{n=1}^\infty E\{Y_n^2\} < \infty \tag{6}$$

and
$$E\{Y_n|X_1, ..., X_n\} = 0 \tag{7}$$
with probability one for all n, imply
$$P[\lim_{n\to\infty} X_n = \theta] = 1. \tag{8}$$

Proof. Without loss of generality we may take $\theta = 0$.

1. From (4) and (6) it follows that $E(X_n^2) < \infty$ for every n.

2. Let $s(n)$ be the sign of $[T_n(X_1, ..., X_n)]\,[X_n]$ if neither factor is zero, and $s(n) = 1$ if either factor is zero. Write
$$\Pi(m, n) = \prod_{j=m}^{n} s(j), \quad Y'_n = \Pi(1, n)\,Y_n.$$
Then $\sum_1^\infty Y'_n$ converges with probability one by lemma [Appendix 2, §4] and (6) and (7). Write
$$Z(m, n) = \sum_{j=m}^{n} Y'_j.$$
For any δ and ϵ both positive, there exists $M_0(\delta, \epsilon)$ such that
$$P\left\{\sup_{\substack{m,\,n\\ M\leqslant m\leqslant n}} |Z(m,n)| > \frac{\delta}{48}\right\} < \epsilon/2. \tag{9}$$

3. Let $\quad d(m, m-1) = 1,$
$$d(m, n) = \prod_{j=m}^{n} (1+\beta_j) \quad \text{for} \quad n \geqslant m.$$
Consider the sum
$$S(m, n) = \sum_{j=m}^{n+1} d(j, n)\,Y'_{j-1}$$
which is equal to
$$\sum_{j+m}^{n-1} Z((m-2), (j-1))\,[d(j, n) - d(j+1, n)] - Y'_{m-2}d(m, n)$$
$$+ Z((m-2), (n-1))\,d(n, n) + Y'_n. \tag{10}$$
Since $d(j, n) \geqslant d(j+1, n)$ we see that the absolute value of (10) is not greater than
$$2\left[\sup_{\substack{j\\ m-1\leqslant j\leqslant n}} |Z((m-2), (j-1))|\right](d(m, n)) + |Y_n|.$$

Hence from (2) and (9) it follows that for δ and ϵ both positive there exists an $M_{00}(\delta,\epsilon) \geqslant M_0(\delta,\epsilon)$ such that $d(m,\infty) < 3/2$ for $m \geqslant M_{00}$ and

$$P\left\{\sup_{\substack{m,\,n\\ M_{00}\leqslant m\leqslant n}} |Z(m,n)| < \delta/48, \sup_{\substack{m,\,n\\ M_{00}\leqslant m\leqslant n}} |S(m,n)| < \delta/8\right\} > 1-\epsilon/2. \tag{11}$$

Hence the result.

THEOREM 3 (Dvoretzky). Let

$\{\alpha_n(X_1,\ldots,X_n)\}, \{\beta_n(X_1,\ldots,X_n)\}$ and $\{\gamma_n(X_1,\ldots,X_n)\}$

be sequences of non-negative functions of real variables $X_1,\ldots,X_n$ such that the functions $\alpha_n(X_1,\ldots,X_n)$ are uniformly bounded and $\lim_{n\to\infty}\alpha_n(X_1,\ldots,X_n) = 0$ uniformly for all sequences $X_1,\ldots,X_n$; (1)

the functions $\beta_n(X_1,\ldots,X_n)$ are measurable and $\sum_1^\infty \beta(X_1,\ldots,X_n)$ is uniformly bounded and uniformly convergent in $X_1,\ldots,X_n$; (2)

and for any $L > 0$ there exist non-negative $\gamma_n(X_1,\ldots,X_n)$ satisfying

$$\sum_1^\infty \gamma_n = \infty, \tag{3}$$

and

$$|T_n(X_1,\ldots,X_n)-\theta| \leqslant \max\{\alpha_n(X_1,\ldots,X_n), [1+\beta_n(X_1,\ldots,X_n)]\,|X_n-\theta|-\gamma_n\} \tag{4}$$

holds uniformly for all sequences $X_1,\ldots,X_n$ for which

$$\sup_{n=1,2\ldots} |X_n| < L, \text{ where } L \text{ is an arbitrary positive number,} \tag{5}$$

where $T_n(X_1,\ldots,X_n)$ are measurable transformations such that

$$X_{n+1} = T_n(X_1,\ldots,X_n)+Y_n(X_n) \quad \text{for} \quad n \geqslant 1 \tag{6}$$

and

$$E(X_1^2) < \infty, \tag{7}$$

$$\sum_1^\infty E(Y_n^2) < \infty; \tag{8}$$

with probability one $E\{Y_n|X_1, ..., X_n\} = 0.$ (9)

Then
$$P\{\lim_{n\to\infty} X_n = \theta\} = 1 \tag{10}$$
and
$$\lim_{n\to\infty} E(X_n - \theta)^2 = 0. \tag{11}$$

Proof. Again, take $\theta = 0$ and let ϵ and δ be positive and arbitrary. To prove (10) it is sufficient to prove that

$$P\{|X_n| < \delta \quad \text{for all } n \text{ sufficiently large}\} > 1 - \epsilon. \tag{12}$$

Let $M \geqslant M_0(\delta, \epsilon)$ be so large that, for $n \geqslant M$, $\alpha_n < \delta/8$. Let L be so large that $L > \delta$ and

$$\operatorname*{Max}_{1\leqslant j\leqslant m} E\{X_j^2\} < \frac{\epsilon L^2}{32M}. \tag{13}$$

We take this to be the L for which (3) holds. It also follows that

$$P\{\operatorname*{Max}_{1\leqslant j\leqslant M} |X_j| \leqslant L/4\} > 1 - \epsilon/2. \tag{14}$$

Suppose that the following four conditions are fulfilled:

$$\text{the relations (11) of Theorem 1,} \tag{15}$$

$$|X_m| \leqslant \delta/4 \quad \text{for some} \quad m \geqslant M, \tag{16}$$

$$|X_{m+1}| > \delta/4,\ 1 \leqslant j \leqslant k, \tag{17}$$

$$|X_{m+k+1}| \leqslant \delta/4, \tag{18}$$

Here $1 \leqslant k \leqslant \infty$. When $k = \infty$, (17) holds for $j \geqslant 1$ and (18) is vacuous (it will be clear by the time the proof is finished that k cannot $= \infty$). Because $\alpha_n < \delta/8$ for $n \geqslant M$ and because of (15), (16) and (17) it follows that

$$|T_{m+j}(X_{m+j})| > \alpha_{m+j} \quad (0 \leqslant j \leqslant k-1). \tag{19}$$

$$\operatorname{sign} X_{m+j+1} = \operatorname{sign} T_{m+j}(X_{m+j}) \quad (0 \leqslant j \leqslant k-1). \tag{20}$$

Applying 4 (with γ's zero) we obtain that X_{m+1} lies between zero and

$$s(m)\,(1 + \beta_m)\,X_m + Y_m. \tag{21}$$

Repeating this argument, one obtains, for $1 \leqslant j \leqslant k$, X_{m+j} lies between 0 and

$$s(m+j-1)\,s(m+j-2)\ldots s(m)\,d(m,m+j-1)\,X_m$$
$$+s(m+j-1)\ldots s(m+1)\,d(m+1,m+j-1)\,Y_m$$
$$+s(m+j-1)\,d(m+j-1,m+j-1)\,Y_{m+j-2}+Y_{m+j-1}. \quad (22)$$

The absolute value of (22) is not greater than

$$|X_m|d(m,m+j-1)+|S(m+1,m+j-1)|. \quad (23)$$

Hence
$$|X_{m+j}| < \delta, \quad 1 \leqslant j \leqslant k. \quad (24)$$

To prove (12) it remains only to show that the following conditions can not both hold:

the relations (14) and (11) of theorem (15), (25)

$$|X_n| > \delta/4 \quad \text{for all} \quad n \geqslant M. \quad (26)$$

Applying the argument as we did before with δ replaced by L we obtain that

$$|X_n| < L \quad \text{for all} \quad n \geqslant 1. \quad (27)$$

Hence (3) holds. In view of (25) and (26) it follows that

$$|T(X_n)| > \alpha_n \quad \text{for all} \quad n \geqslant M-1, \quad (28)$$

$$\operatorname{sign} T_n(X_n) = \operatorname{sign} X_{n+1} \quad \text{for all} \quad n \geqslant M-1. \quad (29)$$

We may now, and do, apply the argument which led to (23), but with the γ's which satisfy (3), and we conclude that for all $n > M$, the absolute value of $|X_n|$ is not greater than

$$|X_M|\,d(M,n-1)+|S(M+1,n-1)|-\sum_{j=M}^{n-1}\gamma_j \quad (30)$$

for n sufficiently large this becomes negative, contradicting (28) and hence (26). This completes the proof of (10).

Let $K = \max_{I \leqslant j < \infty} \alpha_j$. Let N be an integer to be chosen later. In view of (10) we have only to prove that $\lim_{n\to\infty} E\{((|X_n|-k)^+)^2\} = 0$. Let P denote probability measure and A be any set in the sample space which can be defined in terms of $X_1, \ldots, X_m$. We use the inequality

$$
\begin{aligned}
H_{m+1}(A) &= \int_A ((|X_{m+1}| - k)^+)^2 \, dP \\
&= \int_A ((|T_m(X_m) + Y_m| - k)^+)^2 \, dP \\
&\leqslant \int_A [Y_m^2 + ((|T_m(X_m)| - k)^+)^2] \, dP \\
&\leqslant \int_A [Y_m^2 + k\beta_m(1 + k\beta_m) + (1 + \beta_m)^2 (1 + k\beta_m) \\
&\qquad \times ((|X_m| - k)^+)^2] \, dP \qquad (31)
\end{aligned}
$$

which can be deduced from (4) and (9). Let $B(j)$ be the set $\{|X_{N+j}| \leqslant k, |X_{N+i}| > k \text{ for } 0 \leqslant i < j\}$, $D(j)$ the complement of $B(0) + B(1) + \ldots + B(j)$. Iterate the inequality (31) to obtain an upper bound on $H_n(A)$, $n > N$, beginning the iteration at $m = N$, $N+1, \ldots, n-1$, respectively, and using as A the sets

$$B(0), B(1), \ldots, B(n-N-1)$$

respectively. In each case the last term of the integrand of the right number of (31) vanishes. Adding, we obtain that

$$H_n(B(0) + \ldots + (n-N))$$

can be made arbitrarily small by making N sufficiently large.

It remains only to consider $H_n(D(n-N))$. For any point in $D(n-N)$ we have, as in (22), that

$$|X_n| \leqslant |\pi(1, N-1) \, d(N, n-1) \, X_N + S(N+1, n-1)|. \qquad (32)$$

Hence, by Minkowski's inequality

$$
\left(\int_{D(n-N)} (X_n^2) \, dP\right)^{\frac{1}{2}} \leqslant [d(1, \infty)] \left(\int_{D(n-N)} (X_N^2) \, dP\right)^{\frac{1}{2}} + [d(1, \infty)] \left(\sum_{j=N}^{\infty} EY_j^2\right)^{\frac{1}{2}}. \qquad (33)
$$

The second term on the right of (33) can be made arbitrarily small by making N sufficiently large. The first term can be made arbitrarily small by making n sufficiently large, since

$$P\{D(n-N)\} \to 0 \quad \text{as} \quad n \to \infty.$$

This completes the proof of (11).

4. Approximation for asymptotically regular process

In this section a restricted sequence of random variables is considered, but the conditions on iterative coefficients are relaxed. The condition of type $\sum_{1}^{\infty} \beta_n < \infty$ is no longer required.

Theorem 4 asserts the convergence of a sequence of estimates $\{X_n\}$ to θ in mean square and with probability one, and it also shows that the convergence does not depend on the initial elements of the sequence $\{a_n\}$, but it does depend on the final elements. The mathematical content of Theorem 4(a) and 4(b) is the same as that of Comer [15] and the mathematical content of Theorem 4(c) is similar to that of Burkholder [10] though under different conditions. Furthermore it is proved that the conditions are sufficient.

THEOREM 4(a) (Comer). *Let $\{X_n\}$ be a sequence as defined below and X_1 a random variable such that $E(X_1-\theta)^2 < V^2 < \infty$, where θ and V are real numbers.* Suppose that

(i) $X_{n+1} = X_n - a'_n[Y_n(X_n) - Y_0]$, where Y_0 is any real number and $Y_n(X_n)$ is random variable such that $E[Y_n(X)|X_n] = M(X_n)$.

(ii) $L \leqslant d_n = \dfrac{M(X_n)-Y_0}{X_n-\theta} \leqslant u$ for every n, where L and u are real numbers such that $L < u$. Without loss of generality assume $Y_0 = 0$.

(iii) $\sum_{1}^{\infty} a'_n = \infty$, where $\{a'_n\}$ is positive number sequence.

(iv) $\lim_{n\to\infty} a'_n = a' \geqslant 0$.

(v) $0 \leqslant a' \leqslant \dfrac{1}{u}$.

(vi) $Z_n = Y_n(X_n) - M(X_n)$ and M is continuous at θ with $M(0) = 0$.

(vii) $E[Z_n^2] = k^2$, where k is any positive real number. Then

$$\lim_{n\to\infty} [E(X_n - \theta)^2]^{\frac{1}{2}} \leqslant k/L \quad \text{and} \quad \lim_{n\to\infty} \sup [E[Y_n^2]]^{\frac{1}{2}} \leqslant \frac{uk}{L} + k.$$

Proof. From (iv) and (v) it follows that there exists an N such that

$$a_n' u < 1 \quad \text{for} \quad n > N.$$

Therefore

$$\left.\begin{aligned}
0 < 1 - a_n' u \leqslant 1 - a_n' d_n \leqslant 1 - a_n L < 1 \quad &\text{for} \quad n > N, \\
X_{n+1} - \theta = X_n - \theta - a_n'[Z_n] - a_n' d_n (X_n - \theta), \\
X_{n+1} - \theta = (1 - a_n' d_n)(X_n - \theta) - a_n' Z_n \quad (n \geqslant 1), \\
|X_{n+1} - \theta| \leqslant (1 - a_n' L)|X_n - \theta| + a_n' |Z_n|, \\
E|X_{n+1} - \theta|^2 \leqslant (1 - a_n' L)^2 E(X_n - \theta)^2 + a_n'^2 E(Z_n^2) \\
+ 2(1 - a_n' L) a_n' E\{|Z_n||X_n - \theta|\} \\
\leqslant (1 - a_n' L)^2 E(X_n - \theta)^2 + a_n^2 E(Z_n^2) \\
+ 2(1 - a_n' L)(a_n')[EZ_n^2]^{\frac{1}{2}} [E(X_n - \theta)^2]^{\frac{1}{2}}, \\
[E(X_{n+1} - \theta)^2]^{\frac{1}{2}} \leqslant (1 - a_n' L)[E(X_n - \theta)^2]^{\frac{1}{2}} + a_n' k \\
= [E(X_n - \theta)^2]^{\frac{1}{2}} - a_n' L\{[E(X_n - \theta)^2]^{\frac{1}{2}} \\
- k/L\}, n \geqslant N.
\end{aligned}\right\} \quad (1)$$

Let $\epsilon > 0$ then let us assume contrary to the hypothesis (to be proved) that

$$E(X_n - \theta)^2 \geqslant k/L + \epsilon. \quad \text{Let} \quad n > N.$$

Substituting in (1) we shall have

$$[E(X_{n+1} - \theta)^2]^{\frac{1}{2}} \leqslant [E(X_n - \theta)^2]^{\frac{1}{2}} - a_n' L\epsilon,$$

then

$$\left.\begin{aligned}
\sum_N^{n(\epsilon)-1} [E(X_{n+1} - \theta)^2]^{\frac{1}{2}} \leqslant \sum_N^{n(\epsilon)-1} [E(X_n - \theta)^2]^{\frac{1}{2}} - L\epsilon \sum_N^{n(\epsilon)-1} a_n', \\
[E(X_{n(\epsilon)} - \theta)^2]^{\frac{1}{2}} \leqslant [E(X_N - \theta)^2]^{\frac{1}{2}} - L\epsilon \sum_N^{n(\epsilon)-1} a_n'.
\end{aligned}\right\} \quad (2)$$

But since $\sum_1^\infty a'_n = \infty$ one has

$$L\epsilon \sum_{N}^{n(\epsilon)-1} a'_n \geqslant [E(X_N-\theta)^2]^{\frac{1}{2}} - k/L \tag{3}$$

and substituting (3) in (2) one can obtain

$$[E(X_{n(\epsilon)}-\theta)^2]^{\frac{1}{2}} \leqslant k/L.$$

Since $\quad Y_n = d_n(X_n-\theta)+Z_n \quad (n \geqslant 1),$

$$|Y_n| \leqslant u|X_n-\theta|+|Z_n|$$

and $\quad [EY_n^2]^{\frac{1}{2}} \leqslant u[E(X_n-\theta)^2]^{\frac{1}{2}}+k \leqslant uk/L+k.$

THEOREM 4(*b*) (Comer). *Assume conditions* (i) *to* (vi) *of Theorem* 4(*a*)

(i) *Let* $\quad \mu_{n,m} = E(Z_n|Z_{n-m}, Z_{n-m-1}, \ldots, Z_1),$

where $\quad m \geqslant 1 \quad$ *and* $\quad n \geqslant m+1.$

(ii) *There exists a sequence of real numbers* $\{\xi_m\}$ *such that*

$$[E(\mu_{n,m}^2)]^{\frac{1}{2}} \leqslant \xi_m, \quad m \geqslant 1 \quad \textit{and} \quad n \geqslant m+1 \quad \textit{and} \quad \lim_{m\to\infty} \xi_m = 0.$$

Then there exists a function g of a′ such that

$$\limsup_{n\to\infty} E(X_n-\theta)^2 \leqslant g(a'),$$

$$\lim_{n\to\infty} E[(Y_n-Z_n)]^2 \leqslant u^2 g(a')$$

and $\quad g(a') \to 0 \quad \text{as} \quad a' \to 0,$

$$g(0) = 0.$$

Proof. Let $\theta = 0$ without loss of generality. Then

$$\begin{aligned} E(X_{n+1}^2) &\leqslant (1-a'_n L)^2 E(X_n^2) - 2a'_n E[(1-a'_n d_n) X_n Z_n] + a_n'^2 E(Z_n^2) \\ &\leqslant (1-a'_n L)^2 E(X_n^2) - 2a'_n E(X_n Z_n) + 2a_n'^2 E[d_n X_n Z_n] \\ &\qquad + a_n'^2 k^2 \\ &\leqslant \left\{(1-2a'_n L) E(X_n^2) + a_n'^2 L\left(\frac{k}{L}+\epsilon\right)^2\right\} \left\{2a'_n|E(X_{n-m}Z_n)| \right. \\ &\qquad \left. + 2a'_n|E(X_n - X_{n-m}) Z_n]|\right\} + \left\{2a_n'^2 u\left(\frac{k}{L}+\epsilon\right) k + a_n'^2 k^2\right\}, \end{aligned}$$

$$|E(X_{n-m}Z_n)| \leqslant |E[E(X_{n-m}Z_n|Z_{n-m-1}, \ldots, Z_1, X_1)]|$$
$$= |E[X_{n-m}E(Z_n|Z_{n-m-1}, \ldots, Z_1)]|$$
$$\leqslant \left(\frac{k}{L}+\epsilon\right)\xi_{m+1} \quad (m \geqslant 0),$$

$$|X_{n+1}-X_n| \leqslant a'_n|Y_n| \quad \text{for} \quad n \geqslant 1.$$

Now

$$|E[(X_n - X_{n-m})Z_n]| \leqslant \sum_{i=n-m}^{n-1} |E[(X_{i+1}-X_i)Z_n]|$$
$$\leqslant \sum_{i=n-m}^{n-1} [E(X_{i+1}-X_i)^2]^{\frac{1}{2}}k$$
$$\leqslant \sum_{i=n-m}^{n-1} a'_i[E(Y_i^2)]^{\frac{1}{2}}k$$
$$\leqslant k\left[u\left(\frac{1}{L}+\epsilon\right)+k\right]\sum_{i=n-m}^{n-1} a'_i,$$

$$EX_{n+1}^2 \leqslant E(X_n^2) - 2a_n^2L\left\{E(X_n^2) - \frac{1}{2L}\left[2\left(\frac{k}{L}+\epsilon\right)\xi_{m+1} + 2k\left\{u\left(\frac{k}{L}+\epsilon\right)+k\right\}\sum_{i=n-m}^{n-1} a'_i + a'_n\left\{L^2\left(\frac{k}{L}+\epsilon\right)^2 + 2u\left(\frac{k}{L}+\epsilon\right)k + k^2\right\}\right]\right\}.$$

Let $a' > 0$ and let

$$g(a') = \min_m \left\{\frac{1}{2L}\left[2\frac{k}{L}\xi_{m+1} + 2Kma'\left(u\frac{k}{L}+K\right) + a'\left\{L^2\left(\frac{k}{L}\right)^2 + 2uk\frac{k}{L} + k^2\right\}\right]\right\}$$

$$= \min_m \left\{\frac{k}{2L}\left[\frac{2\xi_{m+1}}{L} + \frac{2ma'}{L}k(u+L) + \frac{akL}{L} + a'\frac{(2u+L)k}{L}\right]\right\}$$

$$= \min_m \frac{k}{2L^2}[2\xi_{m+1} + 2ma'K(u+L) + a'KL + a'(2u+L)K].$$

Hence there exists δ such that for $n > n(\delta)$

$$E(X_{n+1}^2) \leqslant E(X_n^2) - 2a'_nL\{(X_n^2) - g(a') - \delta/2\}$$

if $$E(X_n^2) > g(a')+\delta$$

then $$E(X_{n+1}^2) \leqslant E(X_n^2) - 2a_n' L\delta/2,$$

$$\sum_{n_0(\delta)}^{n_1(\delta)-1} E(X_{n+1}^2) \leqslant \sum_{n_0(\delta)}^{n_1(\delta)-1} E(X_n^2) - L\delta \sum_{n_0(\delta)}^{n_1(\delta)-1} a_n',$$

$$E(X_{n_1(\delta)}^2) \leqslant E(X_{n_0(\delta)}^2) - L\delta \sum_{n_0(\delta)}^{n_1(\delta)-1} a_n'$$

but $$L\delta \sum_{n_0(\delta)}^{n_1(\delta)} a_n' \geqslant E(X_{n_0(\delta)}^2) - g(a'),$$

this implies that $$E(X_{n_1(\delta)}^2) \leqslant g(a').$$

The other results follow immediately. Hence the theorem is proved.

THEOREM 4(c). *Assume that the conditions of Theorem* 4(*a*) *and* 4(*b*) *hold true. In addition*

(i) $a_n' \to 0$ as $n \to \infty$
(ii) $(X_n - \theta) M(X_n) > 0$.

Then, $$P[\lim_{n\to\infty} X_n = \theta] = 1.$$

Proof. Let $\{X_n\}$ be the sequence of random variables under consideration. Then $\sum_1^n \{(X_i-\theta) - E((X_i-\theta)|X_1, \ldots, X_{n-1})\}$ converges to a random variable, as $n \to \infty$ with probability one, because $\sum_1^\infty E(X_n-\theta)^2 < \infty$ from Theorem 4(*b*).

$$X_{i+1} - \theta = X_i - \theta - a_i' Y_i,$$

$$\begin{aligned} E(X_{i+1}-\theta|X_1, \ldots, X_{i-1}X_i) &= X_i - \theta - a_i' E(Y_i|X_1, \ldots, X_i) \\ &= X_i - \theta - a_i' M(X_i), \end{aligned}$$

$$\sum_1^{n+1} [(X_i-\theta) - E(X_i - \theta|X_1, \ldots, X_{i-1})] = X_{n+1} - X + \sum_1^n a_i' M(X_i),$$

$X_{n+1} + \sum_1^n a_i' M(X_1)$ converges to a random variable with probability one.

Let $$v_1 = \{w \mid \liminf_{n\to\infty} X_n(w) = \infty\} = \{\lim_{n\to\infty} X_n = \infty\},$$

$$v_2 = \{\liminf_{n\to\infty} X_n = -\infty\},$$

$$v_3 = \{\limsup_{n\to\infty} X_n < \limsup_{n} X_n\},$$

$$w = \left\{X_{n+1} + \sum_1^n a'_j M(X_j) \text{ converges}\right\}.$$

Let $P(wv_1) > 0$. Let $\omega \in wv_1$ and consider the real number sequence $\{X_n(\omega)\}$. Denote $X_n(\omega)$ by X_n. Thus the real number sequence $\{X_n\}$ satisfies $\liminf_{n\to\infty} X_n = \infty$, $X_{n+1} + \sum_1^n a'_j M_j(X_j)$ converges. Let $P_1 < n$ such that $X_n > 1+\theta$ and let P_2 such that $m > n > p_2$ then

$$X_m + \sum_1^{m-1} a'_j M_j(X_j) - X_n - \sum_{j=1}^{n-1} a'_j M_j(X_j) < 1$$

or $$X_m - X_n + \sum_n^{m-1} a'_j M'_j(X_j) < 1.$$

Thus if $p = p_1 + p_2 + Q(1)$ we have $m > p$

$$X_m \leqslant 1 + X_p - \sum_{j=p}^{m-1} a'_j M_j(X_j)$$

$$\leqslant 1 + X_p, a'_j > 0, M_j(X_j) > 0,$$

because $$X_j > 1+\theta, \quad j > p > M(1).$$

Thus $\{X_n\}$ is bounded on the right contradicting the fact that $\lim_{n\to\infty} X_n = \infty$. Thus $P[wv_1] = 0$. Similarly $P[wv_2] = 0$. Let us prove now $P[wv_3] = 0$.

Suppose $p[wv_3] > 0$. Let $\omega \in wv_3$. Let $X_n = X_n(\omega)$. Then $X_{n+1} + \sum_{j=1}^n a'_j M_j(X_j)$ converges and $\liminf_{n\to\infty} X_n < \limsup_{n\to\infty} X_n$.

Consider the first of the two cases

$$\text{(i) } \theta < \limsup_n X_n; \qquad \text{(ii) } \liminf_{n\to\infty} X_n < \theta.$$

There are real numbers a and b such that

$$\theta < a < b < \limsup_{n\to\infty} X_n \quad \text{and} \quad |M(X)| < u|X|$$

for all n, X. There is a positive integer n and a $p > M(a/2)$ such that if $m > n > p$ then

$$X_m - X_n + \sum_{j=1}^{m-1} a_j' M_j(X_j) < \frac{b-a}{6} \quad \text{and} \quad a_n' < \min\left[\frac{1}{3u}, \frac{b-a}{6u|a|}\right].$$

There are numbers r and s such that $p < r < s$, $X_r < a$, $b < X_s$ and if $s-r > 1$ then $a \leqslant X_j \leqslant b$ implying $M_j(X_j) > 0$ for $r < j < s$. Clearly $X_s - X_r > b-a$ and

$$|X_r| \leqslant |X_r - a| + |a| = a - X_r + |a| < X_s - X_r + |a|$$

$$X_s - X_r + \sum_{j=1}^{s-1} a_j' M_j(X_j) < \frac{b-a}{6}$$

and
$$X_s - X_r \leqslant \frac{b-a}{6} - \sum_{j=r}^{s-1} a_j' M_j(X_j)$$

since
$$a_j' > 0, \quad M_j(X_j) > 0, \quad j > r$$

$$\leqslant \frac{b-a}{6} - a_r' M_j(X_j)$$

$$\leqslant \frac{b-a}{6} + a_r u |X_i|$$

$$\leqslant \frac{b-a}{6} + a_r u [X_s - X_r + |a|]$$

$$X_s - X_r < \frac{b-a}{2},$$

a contradiction because $X_s - X_r > b-a$. Hence $P[wv_3] = 0$. Similarly one can consider case (ii) $\liminf_{n\to\infty} X_n < \theta$. Hence $P[\liminf_{n\to\infty} X_n = \limsup_{n\to\infty} X_n = X_0] = 1$, and so

$$P(w, \lim_{n\to\infty} X_n = X_0, X_0 = \theta) + P(w, \lim_{n\leftarrow\infty} X_n = X_0, X_0 > \theta)$$
$$+ P(w, \lim_{n\to\infty} X_n = X_0, X_0 < \theta) = 1.$$

We shall show that

$$P(w, \lim_{n\to\infty} X_n = X_0, X_0 > \theta) = 0$$

and $$P(w, \lim_{n\leftarrow\infty} X_n = X_0, X_0 < \theta) = 0.$$

Let $$P(w, \lim_{n\to\infty} X_n = X_0, X_0 > \theta) > 0.$$

Let

$$0 < \delta_1 < \delta_2 \quad \text{and} \quad P(w, \lim_{n\to\infty} X_n = X_0, \theta+\delta_1 < X_0 < \theta+\delta_2) > 0.$$

Let ω be a point in the above set $\{X_n(\omega)\}$, denoted by $\{X_n\}$ satisfying $X_{n+1} + \sum_1^n a_j' M_j(X_j)$ converges, and $\lim_{n\to\infty} X_n = X_0(\omega)$ $\theta+\delta_1 < X_1(\omega) < \theta+\delta_2$. Thus there is a $p > Q(\delta_1/2)$ such that if $n > p$ then $\theta+\delta_1 < X_n < \theta+\delta_2$ thus for $n > p$

$$X_{n+1} + \sum_{j=1}^{n} a_j' M_j(X_j) > \theta+\delta_1 + \sum_{j=1}^{p} a_j' M_j(X_j) + \sum_{p+1}^{n} a_j' M_j(X_j)$$

$$\geqslant \theta+\delta_1 + \sum_{j=1}^{p} a_j' M_j(X_j) + \sum_{p+1}^{n} a_j' L\delta_1.$$

Since $\sum_{p+1}^{n} a_j' \to \infty$ as $n \to \infty$. Therefore

$$\lim_{n\to\infty} \left[X_{n+1} + \sum_1^n a_j' M_j(X_j)\right] > \infty$$

which contradicts the fact that $X_{n+1} + \sum_1^n a_j' M_j(X_j)$ is convergent almost everywhere to a finite number. Therefore

$$P(w, \lim_{n\to\infty} X_n = X_0, X_0 > \theta) = 0$$

and similarly

$$P(w, \lim_{n\to\infty} X_n = X_0, X_0 < \theta) = 0.$$

Hence $$p[\lim_{n\to\infty} X_n = \theta] = 1.$$

Sufficient condition

The sufficient conditions for conditions (i) and (ii) of Theorem 4(*b*) are as follows. Let the sequence $\{Z_n\}$ be regular with no deterministic component. Following Doob [22], one can write

$$Z_n = \sum_{i=0}^{\infty} c_i \eta_{n-i},$$

where the sequence $\{\eta_n\}$ is an orthonormal sequence of Gaussian random variables defined for all integers n on $(-\infty, \infty)$. Then

$$E(Z_n^2) = \sum_{j=1}^{\infty} c_i^2 \leqslant \xi^2 < \infty.$$

If one defines

$$\mu_{n,m} = E\left[\sum_{i=m}^{\infty} c_i \eta_{n-i} | Z_{n-m}, \ldots, Z_1\right],$$

$$E(\mu_{n,m}^2) \leqslant \sum_{i=m}^{\infty} c_i^2$$

which approaches 0 as $m \to \infty$ as required in conditions (ii) Theorem 4 (*b*).

5. Small sample theory

Since the methods displayed in previous sections will in practice only be applied to problems involving a finite sample size, the question naturally arises as to how well asymptotic theory approximates to the true situation. In mathematics the concept of linearity plays a very important role, for example in the Kakutani extension of the Brauwer fixed point theorem which is in a way related to the problem in hand. Thus it is assumed that $M(X) = E(y(X))$ is a straight line and variance of X is taken to be a constant.

Let X_i be the ith estimate of the root θ of $M(X)$ in the Robbins–Monro method, i.e. choose X_1 arbitrarily and define

$$X_i = X_{i-1} + a_{i-1}(M_0 - Y_{i-1}) \quad (i = 2, 3, \ldots). \tag{1}$$

Then assume that Y_i, the response at X_i, is of the form

$$Y_i = M_0 + b(X_i - \theta) + \epsilon_i \sigma$$

with $\quad M_0 = M(\theta),\ b = M'(\theta)$

and that the ϵ_i's are independent random variables with $E(\epsilon_i) = 0$ and var $(\epsilon_i) = 1$, for all i. Thus

$$\begin{aligned} X_{n+1} &= X_n - a_n[b(X_n - \theta) + \epsilon_n \sigma] \\ &= X_n - a_n b(X_n - \theta) - a_n \epsilon_n \sigma \end{aligned} \tag{2}$$

so that

$$(X_{n+1}-\theta) = (X_n-\theta)-a_n b(X_n-\theta)-a_n\epsilon_n\sigma. \tag{3}$$

Let $\quad V_n = X_n-\theta \quad \text{and} \quad a_n = c/nb$

then it follows that

$$V_{n+1} = V_n(1-c/n)-(c/n)(\sigma/b)\,\epsilon_n. \tag{4}$$

Let $V_1 = \Delta$. It can easily be seen that

$$V_n = \Delta \prod_{k=1}^{n-1}(1-c/k)-(\sigma/b)\sum_{i=1}^{n-1}\left[c/i \prod_{k=i+1}^{n-1}(1-c/k)\right]\epsilon_i. \tag{5}$$

By taking expectations, one finds:

$$E(V_n) = E(X_n-\theta) = \Delta \prod_{k=1}^{n-1}(1-c/k), \tag{6}$$

$$\operatorname{var}(V_n) = \operatorname{var}(X_n) = \sigma^2/b^2 \sum_{i=1}^{n-1}\left[c/i \prod_{k=i+1}^{n-1}(1-c/k)\right]^2. \tag{7}$$

Thus equations (6) and (7) give the exact bias and variance of the estimate X_n at any step n of the estimation process. One can easily prove that

$$\lim_{n\to\infty} E(V_n) = 0, \tag{8}$$

$$\lim_{n\to\infty} \operatorname{var}(V_n) = \frac{\sigma^2 c^2}{nb^2(2c-1)}. \tag{9}$$

Thus equations (6) and (7) for each stage of computation can be compared with the asymptotic results (8) and (9).

6. Various modifications of the Robbins–Monro method

Two stage method

Cochran and Davis [13] in their study of the statistics of bioassay show that one can reduce the mean square error of an estimate by adopting a two stage procedure of stochastic approximation. This is mathematically formulated by Venter [72] who proves the convergence of a sequence of estimates to the desired value with probability one. See problem 1.

Accelerated stochastic approximation

If an estimate of θ is in the neighbourhood of θ, then one expects by the Robbins–Monro method that the estimate fluctuates frequently about θ. If it is not in the neighbourhood, then this is not expected to occur. Kesten [49] exploited this idea and formulated an accelerated stochastic-approximation method, which is stated as problem 3.

Blum and Burkholder suggest their iteration techniques and prove convergence with probability one. Odell [60] suggests basing the coefficients a_k on observed values of y's.

Stochastic approximation in presence of trend

Dupač [28] gives a Robbins–Monro method when the optimum value of θ_n moves during the approximation process. This is stated as problem 6 and can be proved by the standard method.

Fabian modification. Fabian [31] suggests the following modification. The recurrence relation for the Robbins–Monro procedure can be rewritten in the following form.

$$X_{n+1} = X_n + a_n|Y_n - \alpha| \operatorname{sign}[Y_n - \alpha].$$

Hence we see that the direction of the nth move of the approximation process is chosen to be sign $[Y_n - \alpha]$ and the length of the move is chosen to be $a_n|Y_n - \alpha|$. This choice will be reasonable if large values of $|Y_n - a|$ can be expected for large $|X_n - \theta|$, but this is not guaranteed by the assumptions of the Robbins–Monro method. Thus one might consider the following modification.

$$X_{n+1} = X_n - a_n (\operatorname{sign}[Y_n - \alpha]).$$

Instead of multiplying a_n by $(Y_n - \alpha)$ one should multiply by the sign of $(Y_n - \alpha)$. This procedure is helpful provided the estimate X_n is in the vicinity of θ and $|Y_n - \alpha|$ is greater than 1.

7. Problems

1. Let X_1 be an arbitrary random variable with finite variance and let the sequence $\{X_n\}$ be generated by

(i) $X_{n+1} = X_n - b_n[Y_n - \alpha]$, where $\{Y_n\}$ is a sequence of random variables with var $Y_n < \infty$ and $E(Y_n) = M(X_n)$. Let

(ii) $b_n = \begin{cases} a_1 & \text{if} \quad n \leqslant N, \\ a_{k+1} & \text{if} \quad n = N+k, k = 1, 2, \ldots, \end{cases}$

where N is a positive integer-valued random variable defined by $N = m$ if and only if sign of $(Y_j - \alpha) =$ sign of $(Y_{j-1} - \alpha)$ for $j = 1, 2, \ldots, m-1$ and sign of $(Y_m - \alpha) \neq$ sign of $(Y_{m-1} - \alpha)$.

(iii) For some $\theta \in R$ real line and for any $\epsilon > 0$

$$\inf_{\epsilon < x - \theta < \infty} [M(x) - \alpha] > 0 \quad \text{and} \quad \sup_{\epsilon < x - \theta < \infty} M[(x) - \alpha] < 0.$$

(iv) For some constants k_2 and k_3 both positive for some $\theta \in R$ and for all $x \in R$, $|M(X) - \alpha| \leqslant k_2|X - \theta| + k_3$.

(v) $\sup_{x \in R} E[Y(x) - M(x)]^2 < \infty$. Then prove that $X_n \to \theta$ with probability 1 as $n \to \infty$. (Venter)

2. Assume the conditions of the problem 1 and also let

$$\liminf_{t \to 0 |X - \theta| \leqslant t} P[Y(x) - \alpha > 0] > 0 \text{ and } \liminf_{t \to 0 |X - \theta| \leqslant t} [Y(x) - \alpha \leqslant 0] > 0.$$

Then prove that $P[N < \infty] = 1$. (Venter)

3. Let θ be a real number and T_n $(n = 1, 2, \ldots)$ be measurable transformations. Let X_1 and Y_n $(n = 1, 2, \ldots)$ be random variables and $\{a_n\}$ a sequence of positive number and define

$$X_{n+1} = T_n(X_1, \ldots, X_n) + b_n(w) Y_n(X_1, \ldots, X_n). \quad (1)$$

The sequence $\{b_n\}$ is selected in the following way from the sequence $\{a_n\}$.

$$b_1 = a_1, \quad b_2 = a_2, \quad b_n = a_{t(n)}, \quad (2)$$

where
$$t(n) = 2 + \sum_{i=0}^{n} l[(X_i - X_{i-1})(X_{i-1} - X_{i-2})] \quad (3)$$

and
$$l(x) = \begin{cases} 1 & \text{if } x \leqslant 0, \\ 0 & \text{if } x > 0. \end{cases}$$

Thus, every time $(X_i - X_{i-1})$ differs in sign from $(X_{i-1} - X_{i-2})$ we take another a_n.

Let $\alpha_n(X_1, \ldots, X_n)$, $\beta_n(X_1, \ldots, X_n)$, $\gamma_n(X_1, \ldots, X_n)$ be non-negative functions and put

$$\epsilon_N = \sup_{X_k} \sum_{n=N}^{\infty} \beta_n(X_1, \ldots, X_n), \quad (4)$$

$$\rho(\delta) = \inf_{n=1,2,\ldots} \inf_{|X_n - \theta| \geqslant \delta} \frac{\gamma_n(X_1, \ldots, X_n)}{b_n}. \quad (5)$$

If

$$|T_n(X_1,\ldots,X_n)-\theta| \leqslant \begin{cases} (1+\beta_n(X_1,\ldots,X_n))|X_n-\theta| \\ \qquad -\gamma_n(X_1,\ldots,X_n) \\ \quad \text{when} \quad (T_n-\theta)(X_n-\theta) > 0, \\ \alpha_n(X_1,\ldots,X_n) \quad \text{when} \\ \quad (T_n-\theta)(X_n-\theta) \leqslant 0, \end{cases} \tag{6}$$

$$\lim_{t(n)\to\infty} \alpha_n(X_1,\ldots,X_n) = 0 \text{ uniformly, for all sequences } X_1, X_2, \ldots \text{ with } t(n)\to\infty, \tag{7}$$

$$\lim_{n\to\infty} \frac{(X_n-\theta)\,\beta_n(X_1,\ldots,X_n)}{b_n} = 0 \text{ uniformly, for all sequences } X_1, X_2, \ldots, \tag{8}$$

$$\lim_{N\to\infty} \epsilon_N = 0, \tag{9}$$

$$\rho(\delta) > 0 \text{ for every positive } \delta, \tag{10}$$

$$\sum_1^\infty a_n = \infty, \quad \sum_1^\infty a_n^2 < \infty \quad \text{and} \quad a_{n+1} \leqslant a_n, \tag{11}$$

$$E(Y_n|X_1,\ldots,X_n) = 0 \quad \text{and} \quad E(Y_n^2|X_1,\ldots,X_n) \leqslant \sigma^2 \text{ with probability one}, \tag{12}$$

$$\liminf_{n\to\infty} \lim_{t\to 0} \inf_{0<|X_n-\alpha|\leqslant t} P\{T_n(X_1,\ldots,X_n) + b_n y_n \leqslant X_n | X_1 = x_1, \ldots, X_n = x_n\} > 0, \tag{13}$$

$$\liminf_{n\to\infty} \lim_{T\to 0} \inf_{0<|X_n-\theta|\leqslant T} P\{T_n(X_1,\ldots,\ldots,X_n)'' b_n Y_n < X_n | X_1 = x_1. \ldots, X_n = x\} > 0, \tag{14}$$

$$(x_1, \ldots, x_{n-1} \text{ arbitrary.})$$

Then prove that X_n converges to θ with probability one. (Kesten)

4. For each positive integer n let R_n be a function from the real numbers into the real numbers. For each ordered pair n, x, where n is a positive integer and x is a real number, let $Z_n(x)$ be a random variable with distribution function $G_n(.|x)$ such that $E[Z_n(x)] = M_n(X)$. Let $\{a_n\}$ be a positive number sequence. Let X_1 be a random variable and if n is a positive integer let

$$X_{n+1} = X_n - a_n Z_n,$$

where Z_n is a random variable with conditional distribution $G_n(./X_n)$ given $X_1, \ldots, X_n, Z_1, \ldots, Z_{n-1}$. Suppose that

(i) There is a function Q from the positive reals into the positive integers such that if

$$\epsilon > 0, \quad |x-\theta| > \epsilon \quad \text{and} \quad n > Q(\epsilon)$$

then
$$(x-\theta)\, M(x) > 0.$$

(ii) $\sup\limits_{n,x} \dfrac{|M_n(x)|}{1+|x|} < \infty.$

(iii) $\sup\limits_{n,x} V_n(x) < \infty$, where $V_n(x) = \operatorname{var} Z_n(x)$.

(iv) If $0 < \delta_1 < \delta_2 < \infty$ then

$$\sum_n a_n [\inf_{\delta_1 \leqslant |x-\theta| \leqslant \delta_2} M_n(x)] = \infty.$$

(v) $\sum\limits_1^\infty a_n^2 < \infty, \quad \sum\limits_1^\infty a_n = \infty.$

(vi) If n is a positive integer then M_n and V_n are measurable.

Prove that

$$P\{\lim_{n\to\infty} X_n = \theta\} = Z. \qquad \text{(Burkholder)}$$

5. (i) Let $G(.|X)$ be a family of distributions of a random variable $Y(X)$ such that $E(Y(X)) = M(X)$.

(ii) Let

$$X_{i+1} = X_i + C_i(\alpha - Y_i), \quad \text{where}$$

(iii) $C_1 = a_1, \quad C_2 = a_2$

if
$$C_{i-1} = a_k \quad \text{for} \quad k \geqslant 2 \quad \text{then}$$

$$C_i = \begin{cases} a_k, & \text{when} \quad \alpha \notin (Y_i, Y_{i-1}), \\ (X_{i+1} - X_i)/(Y_{i+1} - Y_i), & \text{when} \quad \alpha \notin (Y_i, Y_{i-1}). \end{cases}$$

$$C_{i+1} = \begin{cases} a_k \quad \text{when} \quad C_i = a_k \quad \text{and} \quad \alpha \notin (Y_{i+1}, Y_i), \\ (X_{i+1} - X_i)/(Y_{i+1} - Y_i), \quad \text{when} \quad \alpha \in (Y_{i+1}, Y_i), \\ a_{k+1}, \quad \text{when} \quad \alpha \notin (Y_{i+1}, Y_i) \\ \quad \text{and} \quad C_i = (Y_i - Y_i)/(Y_i - Y_{i-1}). \end{cases}$$

(iv) When a_k is an element of sequence $\{a_k\}$ with the properties:

(a) $a_k > 0 \quad \text{for} \quad k = 1, 2, 3, \ldots$

(b) $a_k > a_{k+1} \quad \text{for} \quad k = 1, 2, 3, \ldots$

(c) $\sum_1^\infty a_j = \infty.$

(d) $\sum_1^\infty a_j^2 < \infty.$

(v) Let $M(X)$ be continuous and $G(.|X)$ be such that

$$P[Y > \alpha | X < \theta] < P[Y > \alpha | X = \theta]$$

and
$$P[Y < \alpha | X > \theta] > P[Y > \alpha | X = \theta].$$

(vi) $M(\theta) = \alpha.$

(vii) $|M(X)| \leqslant c + d|X|$, where c and d are constants.

(viii) $\text{var}(Y(X)) < \infty.$

(ix) $M(X) < \alpha \quad \text{for} \quad X < \theta, \quad M(X) > \alpha \quad \text{for} \quad X > \theta.$

(x) $\text{Inf}\,|M(x) - \alpha| > 0$ for every pair of numbers

$$\delta_1 \leqslant |X - \theta| \leqslant \delta_2$$

$$(\delta_1, \delta_2) \quad \text{with} \quad 0 < \delta_1 < \delta_2 < \infty.$$

Prove that
$$P[\lim_{n\to\infty} X_n = \theta] = 1.$$
(Odell)

6. Let $M(X)$ $-\infty < X < \infty$, be an (unknown) real function. Let θ_n $n = 1, 2, \ldots$ be (known) real numbers the first, θ_1, being the unique root of the equation $M(X) = 0$. Let $M_1(X) = M(X)$. For $n = 1, 2, \ldots$ $M_n(X) = M(X - \theta_n + \theta_1)$ so that θ_n is the unique root of $M_n(X) = 0$. Let a_n, $n = 1, 2, \ldots.$ be positive numbers. Let X_1 be a random variable. Define for $n = 1, 2, \ldots$

$$X_{n+1} = X_n^* - a_n Y_n^*,$$

where $X_n^* = (1 + n^{-1}) X_n$ and Y_n^* is a random variable such that

$$E(Y_n^* | X_1, \ldots, X_n) = M_{n+1}(X_n^*),$$

$$\text{var}(Y_n^* | X_1, \ldots, X_n) \leqslant \sigma^2,$$

$$M(X) < 0 \quad \text{for} \quad X < \theta_1$$

and
$$M(X) > 0 \quad \text{for} \quad X > \theta_1,$$

there exist K_0 and K such that

$$K_0|X-\theta_1| \leqslant |M(X)| \leqslant K_1|X-\theta_1| \quad \text{for} \quad -\infty < x < \infty,$$

$$a_n = a/n^\alpha, a > 0, \tfrac{1}{2} < \alpha < 1,$$

θ_n varies in such a way that,

$$\theta_{n+1}-(1+n^{-1})\,\theta_n = O\,(n^{-\omega}),$$

where $\omega > \alpha$; further $E(X_1^2) < \infty$.

Then prove that $X_n - \theta_n \to 0$ in mean square and

$$E(X_n-\theta_n)^2 = O(n^{-\alpha}) \qquad \text{for} \quad \omega \geqslant \tfrac{3}{2}\alpha$$

$$= O(n^{-2(\omega-\alpha)}) \quad \text{for} \quad \omega < \tfrac{3}{2}\alpha. \qquad \text{(Dupač)}$$

7. Let $\{X_n\}$ be a stochastic approximation process with $H(Y|X)$ a family of distribution function corresponding to the parameter X,

$$M(X) = \int Y\,dH(Y|X), \quad \sigma^2(X) = \int (Y-M(X))^2\,dH(Y|X),$$

let $\{a_n\}$ be positive number sequence such that

$$\sum_1^\infty a_n = \infty, \quad \sum_1^\infty a_n^2 < \infty,$$

and let X_1 be arbitrary number, such that

(i) $|M(X)| \leqslant (L|X|+k)f(X)$, where f is a function which is positive and bounded in any finite interval, and l and k are constants.

(ii) $\sigma^2(X) \leqslant \sigma^2 f^2(X)$, where $\sigma^2 < \infty$.

(iii) If $X < X_0$, then $M(X) < 0$, while $X > X_0$, then $M(X) > 0$.

(iv) For every $\delta_2 > \delta_1 > 0 \inf |M(X)| \geqslant 0$

$$\delta_1 \leqslant |X-X_0| < \delta_2$$

and $$X_{n+1} = X_n - a_n Y_n/f(X_n) \quad \text{for} \quad n \leqslant 1.$$

Then prove that $X_n \to X_0$, with probability one and in mean square. (Friedman)

8. Replace condition (i) and (ii) of the problem 7 by

$$\text{(i)}\ |M(X)| \leqslant L|X|+k, \quad \text{(ii)}\ \sigma^2(X) \leqslant \sigma^2 < \infty$$

and add the following conditions

(v) If $X < X_0$ then $M(X) = 0$, while if

$$X > 0, \quad \text{then} \quad M(X) > 0.$$

(vi) For every $0 < \delta$, $\inf\limits_{\delta \leqslant X - X_0 < \infty} |M(X)| > 0$ and choose a_i, δ_i such that:

$$a_i > 0, \quad \sum_1^\infty a_i = \infty, \quad \sum_1^\infty a_i^2 < \infty,$$

$$\delta_i > 0, \quad \Sigma a_i \delta_i = \infty, \quad \delta_i \to 0 \quad \text{as} \quad i \to \infty$$

and define $$X_{n+1} = X_n + a_n(\delta_n - Y_n).$$

Then prove that $X_n \to X_0$ with probability one and in mean square. (Friedman)

9. Prove the result of Theorem 2 by using Lemmas 8 and 9 of Appendix 3. (Derman and Sacks)

10. Under the assumptions of Theorem 2 of Chapter 6 (page 101) prove that $X_n \to \theta$ with probability one.

CHAPTER 3

THE KIEFER–WOLFOWITZ METHOD

1. Introduction

Kiefer and Wolfowitz [50] give a method of determining the location of a maximum of a regression function, (if it exists) and establish asymptotic properties of the procedure. Blum [4] proves under weak conditions that their method converges with probability one. We discuss in this chapter a method of Dupač [27] to determine location of maximum and this method has application in Chapter 7 also.

Among the class of iterating coefficients sequences the optimum sequence of coefficient is singled out. This was first studied by Chung [12] but we follow Dupač [27]. A method of locating an inflection point is given. A general class of stochastic-approximation processes is defined which represents a unified treatment of the subject. An illustrative example of a simple regression model is discussed and it is shown how one can sequentially obtain an estimate of a parameter and compute its mean square error.

2. Stochastic approximation to locate maximum of regression function

THEOREM 1 (Dupač). *$\{H(Y(x)\}$ is a family of distribution functions depending on the real parameter x. $M(x) = \int Y\,dH(Y|x)$ is increasing for $x < \theta$ and decreasing for $x > \theta$.*

For every X

$$\text{I} \qquad \sigma^2(X) = \int [Y - M(X)]^2\,dH(Y|X) \leqslant \sigma^2,$$

$$K_0|X-\theta| \leqslant |M'(X)| \leqslant K_1|X-\theta|,$$

$$\text{II} \qquad (0 < K_0 < K_1 < \infty).$$

Let $\{a_n\}$, $\{c_n\}$ be positive sequence such that

$$c_n \to 0, \quad \sum_1^\infty a_n = \infty, \quad \sum_1^\infty a_n c_n < \infty, \quad \sum_1^\infty a_n^2/c_n^2 < \infty. \tag{1}$$

For a given X_1, define

$$X_{n+1} = X_n + a_n \frac{Y_{2n} - Y_{2n-1}}{c_n},$$

where Y_{2n} and Y_{2n-1} are independently distributed with distributions $H(Y/X_n + C_n)$ and $H(Y|X_n - C_n)$ respectively.

Then X_n converges in mean square to θ.

Proof. For $C > 0$, define

$$M_C(X) = \frac{M(X+C) - M(X-C)}{C}.$$

Let
$$K_\theta(X) = -\frac{M'(X)}{X-\theta}$$

when $X \neq \theta$ and $K_\theta(X) = k_0$ when $X = \theta$. For some $0 < T_i < 1$,

$$M(X+C) = M(X) + M'(X+T_1C)C$$

and
$$M(X-C) = M(X) - M'(X-T_2C)C.$$

Hence

$$\begin{aligned} M_C(X) &= M'(X+T_1C) + M'(X-T_2C) \\ &= -K_\theta(X+T_1C)(X+T_1C-\theta) \\ &\qquad\qquad - K_\theta(X-T_2C)(X-T_2C-\theta) \\ &= -[K_\theta(X+T_1C) + K_\theta(X-T_2C)](X-\theta) \\ &\quad + [T_2K_\theta(X-T_2C) - T_1K_\theta(X+T_1C)]C. \end{aligned}$$

By obvious notation we have

$$M_C(X) \equiv -K_{\theta,C}(X)(X-\theta) + G_{\theta,C}(X)C.$$

From (ii) we have $K_0 \leqslant K_{\theta,C}(X) \leqslant K_1$ for every X, so

$$2K_0 \leqslant K_{\theta,C}(X) \leqslant 2K_1, \quad |G_{\theta,C}(X)| \leqslant K_1. \tag{2}$$

From (2) and

$$(X-\theta)M_C(X) = -K_{\theta,C}(X)(X-\theta)^2 + G_{\theta,C}(X)C(X-\theta)$$

we have

$$\begin{aligned} -2K_1(X-\theta)^2 - K_1C|X-\theta| &\leqslant (X-\theta)M_C(X) \\ &\leqslant -2K_0(X-\theta)^2 + K_1C|X-\theta|. \end{aligned} \tag{3}$$

From $|a+b|^r \leqslant 2^{r-1}(|a|^r+|b|^r)$, $r \geqslant 1$, see page 192 and (2) we also have

$$M_C^2(X) = [-K_{\theta,C}(X)(X-\theta)+G_{\theta,C}(X)C]^2$$
$$\leqslant 2[K_{\theta,C}^2(X)(X-\theta)^2+G_{\theta,C}^2(X)C^2]$$
$$\leqslant 8K_1^2(X-\theta)^2+2K_1^2C^2. \tag{4}$$

From the monotonicity of M, $X < \theta - C$ or $X > \theta + C$ implies $[(X-\theta)\times M_C(X)] < 0$, so $(X-\theta)M_C(X) > 0$ implies $|X-\theta| < c$. Hence, weakening the right side of (3) we have

$$(X-\theta)M_C(X) \leqslant K_1 C|X-\theta| \leqslant K_1 C^2. \tag{5}$$

From the definition of the X_n sequence,

$$(X_{n+1}-\theta)^2 = (X_n-\theta)^2+2a_n(X_n-\theta)Z_n+a_n^2 Z_n^2,$$
$$Z_n \equiv \frac{Y_{n2}-Y_{2n-1}}{C_n}.$$

The assumptions give

$$E\{Z_n|X_n\} = \frac{M(X_n+C_n)-M(X_n-C_n)}{C_n} = M_{C_n}(X_n),$$

$$\operatorname{var}\{Z_n|X_n\} = \frac{\sigma^2(X_n+C_n)+\sigma^2(X_n-C_n)}{C_n^2}.$$

Therefore,

$$E\{(X_{n+1}-\theta)^2|X_n\} = (X_n-\theta)^2+2a_n(X_n-\theta)M_{C_n}(X_n)$$
$$+\frac{a_n^2}{C_n^2}[\sigma^2(X_n+C_n)+\sigma^2(X_n-C_n)]+a_n^2 M_{C_n}^2(X_n). \tag{6}$$

Using (I), (4), and (5), this gives

$$E\{(X_{n+1}-\theta)^2|X_n\} \leqslant (X_n-\theta)^2+2K_1 a_n C_n^2$$
$$+2\sigma^2\frac{a_n^2}{C_n^2}+8K_1^2 a_n^2(X_n-\theta)^2+2K_1^2 a_n^2 C_n^2.$$

Hence,

$$E(X_{n+1}-\theta)^2 \leqslant E(X_n-\theta)^2[1+8K_1^2 a_n^2]$$
$$+\left[2K_1 a_n c_n^2+2\sigma^2\frac{a_n^2}{C_n^2}+2K_1^2 a_n^2 C_n^2\right]$$

or

$$b_{n+1} \leqslant b_n d_n + w_n \quad (n = 1, 2, \ldots).$$

Since $d_n > 0$ for every n we recursively find

$$b_{n+1} \leqslant b_1 \prod_{k=1}^{n} d_k + \sum_{k=1}^{n-1} \left(w_k \prod_{j=k+1}^{n} d_j \right) + w_n.$$

Since $\sum_1^\infty a_n^2 < \infty$, $\Pi d_n < \infty$ and all partial products are uniformly bounded, thus (1) also implies $\Sigma w_n < \infty$. Hence

$$b_n \leqslant B^2 < \infty \tag{7}$$

and
$$E|X_n - \theta| \leqslant E^{\frac{1}{2}}|X_n - \theta|^2 \leqslant B. \tag{8}$$

Going back to (6) and using the right-hand side of (3), rather than (5), to bound $(X_n - \theta) M_{C_n}(X_n)$, there results

$$E\{(X_{n+1} - \theta)^2 | X_n\} \leqslant (X_n - \theta)^2 - 4K_0 a_n (X_n - \theta)^2 + 2K_1 a_n c_n |X_n - \theta| + 2\sigma^2 a_n^2/C_n^2 + 8K_1^2 a_n^2 (X_n - \theta)^2 + 2K_1^2 a_n^2 C_n^2. \tag{9}$$

Apply (7) and (8) to all terms but the first two, after taking expected values, and get

$$E(X_{n+1} - \theta)^2 \leqslant E(X_n - \theta)^2 [1 - 4K_0 a_n] + [2K_1 B a_n C_n + 2\sigma^2 a_n^2/C_n^2 + 8K_1^2 B^2 a_n^2 + 2K_1^2 a_n^2 C_n^2]$$

or
$$b_{n+1} \leqslant b_n d_n' + w_n' \quad (n = 1, 2, \ldots).$$

Since $a_n \to 0$, choose n_0 so that $a_n < (1/4K_0)$, i.e. so that $d_n' > 0$, for all $n \geqslant n_0$. Then as above

$$b_{n+1} \leqslant b_{n_0} \prod_{k=n_0}^{n} d_k' + \sum_{k=n_0}^{n-1} w_k' \prod_{j=k+1}^{n} d_j' + w_n' \quad (n = n_0 + 1, n_0 + 2, \ldots).$$

Since $\sum_1^\infty a_n = \infty$, $\prod_{k=n_0}^{n} d_k'$ converges to zero as $n \to \infty$. Also, from (1) and Kronecker's lemma, $\sum_{n=n_0}^{n-1} w_k' \prod_{j=k+1}^{n} d_j'$ tends to zero as $n \to \infty$.

Hence, for $\epsilon > 0$ arbitrary, there exist n_1 such that

$$\prod_{k=n_0}^{n} d_k' < \frac{\epsilon}{3B^2} \quad \text{for all } n \geqslant n_1, n_2$$

such that

$$\sum_{k=n_0}^{n-1} w_k' \prod_{j=k+1}^{n} d_j' < \epsilon/3 \quad \text{for all } n \geqslant n_2 \quad \text{and} \quad n_3$$

such that $$w'_n < \epsilon/3 \quad \text{for all} \quad n \geqslant n_3.$$
Hence, for all
$$n \geqslant \max(n_1, n_2, n_3) > n_0,$$

$$b_{n+1} \leqslant B^2 \frac{\epsilon}{3B^2} + \frac{\epsilon}{3} + \frac{\epsilon}{3} = \epsilon,$$

i.e. $b_n \to 0$ as $n \to \infty$. Hence the result.

3. Optimum choice of $\{a_n\}$ and $\{c_n\}$

In order to determine optimum choice of $\{a_n\}$ and $\{c_n\}$ one has to study first how fast $b_n = E(X_n - \theta)^2$ goes to zero for sequences of the form
$$a_n = a/n^{\alpha}, \quad c_n = c/n^{\gamma} \quad (a, c > 0), \tag{10}$$
where
$$3/4 < \alpha \leqslant 1, \quad 1-\alpha < \gamma < \alpha - 1/2 \tag{11}$$
are sufficient for (1) to hold. When $\alpha = 1$ we take
$$a > \frac{1}{4K_0}. \tag{12}$$

The following lemmas are due to Chung [12] and are proved in the appendix on inequalities. They will be used in Theorem 2 which determines rate of convergence and Theorem 3 which establishes choice of optimality of $\{a_n\}$ and $\{c_n\}$.

Lemma 1. *If for all* $n \geqslant n_0$
$$b_{n+1} \leqslant \left(1 - \frac{c}{n^s}\right) b_n + \frac{c'}{n^t}, \tag{13}$$
where $0 < s < 1$, $s < t$ *and* $c, c' > 0$ *then*
$$\lim_{n\to\infty} n^{t-s} b_n \leqslant \frac{c'}{c}. \tag{14}$$

Lemma 2. *If for all* $n \geqslant n_0$
$$b_{n+1} \leqslant (1 - c/n)\, b_n + \frac{c'}{n^{p+1}}, \tag{15}$$
where $c > p > 0$ *and* $c' > 0$ *then*
$$b_n \leqslant \frac{c'}{c-p} \cdot \frac{1}{n^p} + 0\left(\frac{1}{n^{p+1}} + \frac{1}{n^c}\right). \tag{16}$$

Both statements remain valid if the inequalities are reversed and lim sup is replaced by lim inf. These are referred to as lemmas 3 and 4. For detailed proof of these lemmas we refer to the appendix on inequalities.

Before proving the theorems we define the following symbols.

DEFINITION. *If $f(n)$ and $g(n) > 0$, $n = 1, 2, \ldots$ are two sequences, then the symbol $f(n) = Og(n)$ means that*

$$\limsup_{n\to\infty} \frac{|f(n)|}{g(n)} < +\infty \quad \textit{(it can be zero)}.$$

DEFINITION. *The symbol $f(n) = O^{-1}(g(n))$ means that*

$$\liminf_{n\to\infty} \frac{|f(n)|}{g(n)} > 0 \quad \textit{(it can be infinity)}.$$

DEFINITION. *The symbol 'o' is used only in connection with*

$$f(n) = o(1), \textit{ i.e.}$$

$$\lim_{n\to\infty} f(n) = 0.$$

THEOREM 2. (Dupač). (I), (II) *of theorem* 1, (10), (11), *and* (12) *imply*

$$b_n = \begin{cases} O\left(\dfrac{1}{n^{2\gamma}}\right) & \textit{when} \quad \gamma < \dfrac{\alpha}{4}, \\ O\left(\dfrac{1}{n^{\alpha-2\gamma}}\right) & \textit{when} \quad \gamma \leqslant \dfrac{\alpha}{4}. \end{cases}$$

(*Note $\gamma < \alpha/4$ can only occur when $\frac{4}{5} < \alpha \leqslant 1$.*)

Take expected values in (9), apply the bound (7) to the next to last term only, and get

$$b_{n+1} \leqslant b_n\left[1 - \frac{4K_0 a}{n^\alpha}\right] + \frac{2K_1 ac}{n^{\alpha+\gamma}} E|X_n - \theta| + \frac{2\sigma^2 a^2}{c^2 n^{2(\alpha-\gamma)}} + \frac{8K_1^2 B^2 a^2}{n^{2\alpha}} + \frac{2K_1^2 a^2 c^2}{n^{2(\alpha+\gamma)}}. \tag{17}$$

For any $\epsilon_n > 0$ we have the inequality

$$\begin{aligned} E|X_n-\theta| &= \int_{|X_n-\theta|\leqslant\epsilon_n} |X_n-\theta|\,dP + \int_{|X_n-\theta|>\epsilon_n} |X_n-\theta|\,dP \\ &\leqslant \epsilon_n P\{|X_n-\theta| \leqslant \epsilon_n\} + \frac{1}{\epsilon_n}\int |X_n-\theta|^2\,dP \\ &\leqslant \epsilon_n + \frac{b_n}{\epsilon_n}. \end{aligned} \tag{18}$$

Let $0 < \epsilon < 4$ and $\eta > 0$ be arbitrary. Put

$$\epsilon_n = \frac{2K_1c}{\epsilon K_0 n^\gamma}.$$

Using (18), (17) becomes

$$\begin{aligned} b_{n+1} &\leqslant b_n\left[1-\frac{4K_0a}{n^\alpha}\right] + \frac{4K_1^2ac^2}{\epsilon K_0 n^{\alpha+2\gamma}} + b_n\frac{\epsilon K_0 a}{n^\alpha} \\ &\qquad + \frac{2\sigma^2a^2}{c^2n^{2\alpha-2\gamma}} + \frac{8K_1^2B^2a^2}{n^{2\alpha}} + \frac{2K_1^2a^2c^2}{n^{2\alpha+2\gamma}} \\ &= b_n\left[1-\frac{(4-\epsilon)K_0a}{n^\alpha}\right] + \frac{4K_1^2ac^2}{\epsilon K_0 n^{\alpha+2\gamma}} \\ &\qquad + \frac{\sigma^2a^2}{c^2n^{2\alpha-2\gamma}}\left[2+\frac{8K_1^2B^2c^2}{\sigma^2n^{2\gamma}} + \frac{2K_1^2c^4}{\sigma^2n^{4\gamma}}\right]. \end{aligned}$$

Hence, for n larger than some $n_0 = n_0(\eta)$, we have

$$\begin{aligned} b_{n+1} &\leqslant b_n\left[1-\frac{(4-\epsilon)K_0a}{n^\alpha}\right] + \frac{4K_1^2ac^2}{\epsilon K_0 n^{\alpha+2\gamma}} + \frac{(2+\eta)\sigma^2a^2}{c^2n^{2\alpha-2\gamma}} \\ &\leqslant b_n\left[1-\frac{(4-\epsilon)K_0a}{n^s}\right] + \frac{1}{n^t}\left[\frac{4K_1^2ac^2}{\epsilon K_0} + \frac{(2+\eta)\sigma^2a^2}{c^2}\right], \end{aligned}$$

where (in the notation of the lemmas)

$$t = \min(\alpha+2\gamma, 2\alpha-2\gamma) = \tfrac{3}{2}\alpha - \tfrac{1}{2}|4\gamma-\alpha| \geqslant 1 \geqslant \alpha = s.$$

For $\alpha < 1$, the conditions of lemma 1 are satisfied with

$$t-s = \tfrac{1}{2}(\alpha-|4\gamma-\alpha|) = \begin{cases} 2\gamma & \text{when } \gamma < \tfrac{1}{4}\alpha, \\ \alpha-2\gamma & \text{when } \gamma \geqslant \tfrac{1}{4}\alpha. \end{cases}$$

Thus $\quad \limsup_{n\to\infty} n^{t-s}b_n \leqslant \dfrac{C'}{C} < \infty \quad$ or $\quad b_n = O\left(\dfrac{1}{n^{t-s}}\right).$

For $\alpha = 1$, we have

$$p = t-1 = \min(1+2\gamma, 2-2\gamma)-1 = \min(2\gamma, 1-2\gamma).$$

Since $0 < \gamma < \frac{1}{2}$,

$$\begin{aligned} 0 < p &\leqslant \max(2\gamma, 1-2\gamma) < 1 \\ &= 1-\tfrac{1}{4}\epsilon \quad \text{for some} \quad 0 < \tfrac{1}{4}\epsilon < 1 \\ &< (1-\tfrac{1}{4}\epsilon)\, 4K_0 a \quad \text{by} \quad (12) \\ &= (4-\epsilon)\, K_0 a = C. \end{aligned}$$

Hence, the conditions of lemma 2 hold, and

$$b_n < \frac{C'}{C-p} O\left(\frac{1}{n^{t-1}}\right) + O\left(\frac{1}{n^t} + \frac{1}{n^{t-1}}\right) = O\left(\frac{1}{n^{t-1}}\right).$$

THEOREM 3. (Dupač). *From Theorem 2 if* (I) *and* (II) *hold,*

$$a_n = \frac{a}{n} \quad \textit{and} \quad c_n = \frac{c}{n^{\frac{1}{4}}}, \quad a > \frac{1}{4K_0} \quad \textit{and} \quad c > 0.$$

then $$b_n = O\left(\frac{1}{\sqrt{n}}\right), \quad \textit{i.e.} \quad \limsup_{n\to\infty} n^{\frac{1}{2}} b_n < \infty.$$

This choice is optimum in the following sense. Suppose

$$a_n = \frac{a}{n^\alpha} \quad \textit{and} \quad C_n = \frac{C}{n^\gamma}$$

with (10), (11), *and* (12) *satisfied. If* $\alpha \neq 1$ *or* $\gamma \neq \frac{1}{4}$ *then there exists an* H, *for which* (I) *and* (II) *hold, such that*

$$b_n = O^{-1}\left(\frac{1}{n^{\frac{1}{2}-\epsilon}}\right), \quad \textit{i.e.} \quad \liminf_{n\to\infty} n^{\frac{1}{2}-\epsilon} b_n > 0 \quad \textit{where} \quad \epsilon > 0.$$

Take $\gamma \geqslant \frac{1}{4}\alpha$ first, and assume

$$\sigma^2(x) \geqslant \sigma_0^2 > 0 \quad \text{for all} \quad x. \tag{19}$$

Using (19) and the left side of (3) in (6), with the last term dropped, we have

$$\begin{aligned} &E\{(X_{n+1}-\theta)^2 | X_n\} \\ &\quad \geqslant (X_n-\theta)^2 - 4K_1 a_n (X_n-\theta)^2 - 2K_1 a_n c_n |X_n-\theta| + 2\sigma_0^2 \frac{a_n^2}{c_n^2}, \end{aligned}$$

so that taking expectations

$$b_{n+1} \geqslant b_n\left[1-\frac{4K_1 a}{n^\alpha}\right]-\frac{2K_1 ac}{n^{\alpha+\gamma}}E|X_n-\theta|+\frac{2\sigma_0^2 a^2}{c^2 n^{2\alpha-2\gamma}}. \tag{20}$$

Substituting $\qquad \epsilon_n = \epsilon\dfrac{\sigma_0^2 a}{2K_1 c^3 n^\gamma} \quad (0 < \epsilon < 2)$

in (18) and using it in (20), we have

$$\begin{aligned} b_{n+1} &\geqslant b_n\left[1-\frac{4K_1 a}{n^\alpha}\right]-\frac{\epsilon a^2\sigma_0^2}{c^2 n^{\alpha+2\gamma}}-\frac{4K_1^2 c^4}{\epsilon\sigma_0^2 n^\alpha}b_n+\frac{2\sigma_0^2 a^2}{c^2 n^{2\alpha-2\gamma}} \\ &\geqslant b_n\left[1-\frac{4K_1 a+4K_1^2 c^4\epsilon^{-1}\sigma_0^{-2}}{n^\alpha}\right]+\frac{(2-\epsilon)\,a^2\sigma_0^2}{c^2 n^{2\alpha-2\gamma}}, \end{aligned}$$

since $\gamma \geqslant \frac{1}{4}\alpha$ entails $\alpha+2\gamma \geqslant 2\alpha-2\gamma$. For $\alpha \neq 1$, the conditions of lemma, 3 are satisfied. Hence

$$\liminf_{n\to\infty} n^{\alpha-2\gamma}b_n \geqslant \frac{C'}{C} > 0,$$

where $\alpha-2\gamma = \frac{1}{2}-\epsilon$ for some $\epsilon > 0$ since $\frac{1}{4}\alpha \leqslant \gamma < \alpha-\frac{1}{2}$ implies $0 < \alpha-2\gamma < \frac{1}{2}$ when $\alpha < 1$. For $\alpha = 1$, the conditions of lemma 4 hold if ϵ is chosen so small that

$$C = 4K_1 a+\frac{4K_1^2 c^4}{\epsilon\sigma_0^2} \quad \text{is} \quad > \tfrac{1}{2} \geqslant 1-2\gamma = p > 0.$$

Hence,

$$n^p b_n \geqslant \frac{C'}{C-p}+O\left(\frac{1}{n}+\frac{1}{n^{C-p}}\right) \quad \text{giving} \quad \liminf_{n\to\infty} n^{1-2\gamma}b_n > 0,$$

where $1-2\gamma = \frac{1}{2}-\epsilon$ for some $\epsilon > 0$ since $\gamma > \frac{1}{4}$. Hence, in the region $\gamma \geqslant \frac{1}{4}\alpha$,

$$b_n = O^{-1}\left(\frac{1}{n^{\frac{1}{2}-\epsilon}}\right) \quad \text{for some} \quad \epsilon > 0$$

provided $\alpha \neq 1$ or $\gamma \neq \frac{1}{4}$. Next take $\gamma < \frac{1}{4}\alpha$, and an H such that

$$M(X) = \begin{cases} -(X)-\theta)^2 & \text{for} \quad X \leqslant 0, \\ -\frac{1}{2}(X-\theta)^2 & \text{for} \quad X > \theta. \end{cases} \tag{21}$$

Then (II) holds with $K_0 = 1$ and $K_1 = 2$. Let (I) be satisfied with $\sigma^2 = 1$, and for simplicity put

$$\theta = 0. \tag{22}$$

Using (4) and (7) in (6),

$$b_{n+1} \leqslant b_n + \frac{2a}{n^\alpha} E\{X_n M_{c_n}(X_n)\} + 2\frac{a^2}{c^2 n^{2\alpha-2\gamma}} + \frac{8K_1^2 a^2 B^2}{n^{2\alpha}} + \frac{2K_1^2 a^2 c^2}{n^{2\alpha+2\gamma}}$$

$$= b_n + \frac{2a}{n^\alpha} E\{X_n M_{c_n}(X_n)\} + \frac{a^2}{c^2 n^{2\alpha-2\gamma}}\left[2 + \frac{8K_1^2 B^2 c^2}{n^{2\gamma}} + \frac{2K_1^2 c^4}{n^{4\gamma}}\right]$$

so that for $n_0 \geqslant n_0(\eta)$, where $\eta > 0$ is arbitrary,

$$b_{n+1} \leqslant b_n + \frac{2a}{n^\alpha} E\{X_n M_{c_n}(X_n)\} + \frac{(2+\eta)\, a^2}{c^2 n^{2\alpha-2\gamma}}. \tag{23}$$

Also from (6)

$$b_{n+1} \geqslant b_n + \frac{2a}{n^\alpha} E\{X_n M_{c_n}(X_n)\}. \tag{24}$$

First we need an inequality on $E\{X_n M_{c_n}(X_n)\}$. From (21) we have for all X and $c > 0$,

$$M_c(X) = \begin{Bmatrix} -4X & \text{for} \quad X \leqslant -c, \\ \dfrac{1}{2c}X^2 - 3X + \tfrac{1}{2}c & \text{for} \quad -c < X < c, \\ -2X & \text{for} \quad X \geqslant c, \end{Bmatrix} \tag{25}$$

$$XM_c(X) = \begin{Bmatrix} -4X^2 \\ \dfrac{1}{2c}X^3 - 3X^2 + \tfrac{1}{2}cX \\ -2X^2 \end{Bmatrix} \leqslant -2X^2 + c^2. \tag{26}$$

If $\chi(a\ b) = 1$ for $a \leqslant X \leqslant b$ and 0 otherwise then (26) can be written

$$XM_c(X) = -4X^2 + X^2\left(1 + \frac{1}{2c}X\right)\chi(-c, c) + \tfrac{1}{2}cX\chi(-c, c) + 2X^2\chi(c, \infty).$$

The second term is positive and the fourth is

$$\geqslant \tfrac{1}{2}X^2\chi(c,\infty) \geqslant \tfrac{1}{2}cX\chi(c,\infty),$$

and hence

$$XM_c(X) \geqslant -4X^2 + \tfrac{1}{2}cX\chi(-c,\infty) \geqslant -4X^2 + \tfrac{1}{2}cX. \tag{27}$$

From (26) and (27) respectively it follows that

$$E\{X_n M_{c_n}(X_n)\} \leqslant -2b_n + \frac{c^2}{n^{2\gamma}}, \tag{28}$$

$$E\{X_n M_{c_n}(X_n)\} \geqslant -4b_n + \frac{c}{2n^\gamma} E(X_n). \tag{29}$$

Substituting (28) in (23) we get

$$b_{n+1} \leqslant b_n\left[1-\frac{4a}{n^\alpha}\right]+\frac{2ac^2}{n^{\alpha+2\gamma}}+\frac{(2+\eta)\,a^2}{c^2 n^{2\alpha-2\gamma}}$$

$$= b_n\left[1-\frac{4a}{n^\alpha}\right]+\frac{ac^2}{n^{\alpha+2\gamma}}\left[2+\frac{(2+\eta)\,a}{c^4 n^{\alpha-4\gamma}}\right].$$

For $\eta' > 0$ we therefore have, since $\gamma < \frac{1}{4}\alpha$,

$$b_{n+1} \leqslant b_n\left[1-\frac{4a}{n^\alpha}\right]+\frac{(2+\eta')\,ac^2}{n^{\alpha+2\gamma}}$$

for all $n \geqslant n_0(\eta')$. From lemmas 1 and 2 we then have, since η' is arbitrary,

$$\limsup_{n\to\infty} n^{2\gamma} b_n \leqslant \begin{cases} \frac{1}{2}c^2 & \text{if} \quad \alpha < 1, \\ \dfrac{2ac^2}{4a-2\gamma} \equiv qc^2 & \text{if} \quad \alpha = 1. \end{cases}$$

Since $\gamma < \frac{1}{4} < a$ (by (12) with $K_0 = 1$) we have $\frac{1}{2} < q < 1$. Hence for any α

$$b_n \leqslant \frac{q'c^2}{n^{2\gamma}} \quad \text{for all,} \quad n \geqslant n_0(q') \quad (q < q' < 1).$$

From Chebyshev's Inequality we thus have

$$P\left\{|X_n| \geqslant \frac{c}{n^\gamma}\right\} \leqslant \frac{b_n n^{2\gamma}}{c^2} \leqslant q', \quad P\left\{|X_n| < \frac{c}{n^\gamma}\right\} \geqslant 1-q' = p' \tag{30}$$

provided $n > n_0(q')$. From the definition of X_n

$$E\{X_{n+1}\} = E\{X_n\}+\frac{a}{n^\alpha}E\{M_{c_n}(X_n)\}. \tag{31}$$

From (25)

$$M_c(X) = -X\chi(-\infty,-c)-3X+\tfrac{1}{2}cX^2\chi(-c,c) + \tfrac{1}{2}c\chi(-c,c)+X\chi(c,c)$$

$$\geqslant -3X+\tfrac{1}{2}c\chi(-c,c)$$

so that, using (30),

$$EM_{c_n}(X_n) \geqslant -3E\{X_n\}+\tfrac{1}{2}c_n P\{|X_n| < c_n\}$$

$$\geqslant -3E\{X_n\}+\frac{cp'}{2n^\gamma}.$$

Substituting in (31) we have

$$E\{X_{n+1}\} \geqslant E\{X_n\}\left[1-\frac{3a}{n^\alpha}\right]+\frac{acp'}{2n^{\alpha+\gamma}}$$

for all $n \geqslant n_0(q')$. Applying lemmas 3 and 4 we have

$$\liminf_n n^\gamma E\{X_n\} \geqslant \begin{cases} \frac{1}{6}cp' & \text{for} \quad \alpha < 1, \\ \dfrac{acp'}{2(3a-\gamma)} > \dfrac{cp'}{6} & \text{for} \quad \alpha = 1 \end{cases}$$

and hence, in either case for $n >$ some n_1,

$$E\{X_n\} > \frac{cp'}{7n^\gamma}. \tag{32}$$

Substituting this in (29) we have

$$E\{X_n M_{c_n}(X_n)\} \geqslant -4b_n + \frac{c^2p'}{14n^{2\gamma}},$$

and this, in turn, in (24) gives for $n > n_1$

$$b_{n+1} \geqslant b_n\left[1-\frac{8a}{n^\alpha}\right]+\frac{ac^2p'}{7n^{\alpha+2\gamma}}.$$

Applying lemmas 3 and 4 again, we get

$$\liminf_n n^{2\gamma} b_n \geqslant \frac{c'p'}{56} > 0,$$

i.e.
$$b_n = O^{-1}\left(\left(\frac{1}{n^{2\gamma}}\right)\right) = O^{-1}\left(\frac{1}{\frac{1}{n^2}-\epsilon}\right) \quad (\epsilon > 0).$$

This establishes optimality of $\{a/n\}$ and $\{c/n^{\frac{1}{4}}\}$ in the class defined by (10) and (11).

4. Class of stochastic-approximation processes

In this section we follow Burkholder [10] and define a class of stochastic-approximation processes. In a way he generalized the method of Kiefer and Wolfowitz of determining location of maximum or minimum. For example one can determine the location of an inflection point if it exists by his method. Also it symbolizes a unified treatment of the subject of stochastic approximation.

DEFINITION. *For each positive integer n let R_n be a function from the real numbers into the real numbers. For each ordered pair (n, x). Where n is a positive integer and x is a real number let $Z_n(x)$ be a random variable with distribution function $G_n(.|x)$ such that*

$$E[Z_n(x)] = R_n(x).$$

Let $\{a_n)$ be a positive number sequence. Let X_1 be a random variable and if n is a positive integer let

$$X_{n+1} = X_n - a_n Z_n,$$

where Z_n is a random variable with conditional distribution

$$G_n(.|X_n) \quad \text{given} \quad X_1, \ldots, X_n Z_1, \ldots, Z_{n-1}.$$

$\{X_n\}$ is called a stochastic-approximation process of the type A_0.

DEFINITION. *Let M be a real-valued function from the real numbers into real numbers. For each real number x let $Y(x)$ be a random variable with distribution function $H(.|x)$ such that $E[Y(x)] = M(x)$. Let $\{a_n\}$ be a positive number sequence, $\{r_n\}$ a positive integer sequence and α real number. Let X_1 be a random variable and for each natural number n let*

$$X_{n+1} = X_n - a_n(Y_n - \alpha),$$

where Y_n is a random variable with conditional distribution function $H^{[r_n]}(.|X_n)$ given $X_1, \ldots, X_n, Y_1, \ldots, Y_{n-1}$. The random variable sequence $\{X_n\}$ will be called a stochastic-approximation process of type A_1.

DEFINITION. *Let M be a real-valued function from the real numbers into real numbers. For each real number x, let $Y(x)$ be a random variable with distribution function $H(.|x)$ such that $E[Y(x)] = M(x)$. Let each of $\{a'_n\}$ and $\{c_n\}$ be a positive number sequence and let $\{r_n\}$ be a positive integer sequence. Let X_1 be a random variable and if n is a positive integer let*

$$X_{n+1} = X_n - \frac{a'_n}{c_n}(Y_{2n-1} - Y_{2n}),$$

where Y_{2n-1} *and* Y_{2n} *are random variables which are conditionally independently distributed according to* $H^{[r_n]}(.\,|X_n-c_n)$ *and* $H^{[r_n]}(.\,|X_n+c_n)$ *respectively, given* $X_1, X_2, \ldots, \ldots, X_n, Y_1, \ldots, Y_{2n-2}$. *The random variable sequence* $\{X_n\}$ *will be called a stochastic approximation process of type* A_2.

DEFINITION. *Let* M *be a real-valued function from the real numbers into real numbers. For each real number* x, *let* $Y(x)$ *be a random variable with distribution function* $H(.\,|x)$ *such that* $E(Y(x)) = M(x)$. *Let* $\{a'_n\}$, $\{c_n\}$ *be positive number sequences, and* $\{r_n\}$ *be a positive integer sequence. Let*

$$X_{n+1} = X_n - \frac{a'_n}{c_n^2}\left[Y_{3n-1} - \frac{Y_{3n-2}+Y_{3n}}{2}\right],$$

where Y_{3n-2}, Y_{3n-1} *and* Y_{3n} *are random variable which are conditionally independently distributed according to*

$$H^{[r_n]}(.\,|X_n-c_n),\quad H^{[r_n]}(.\,|X_n)\quad \text{and}\quad H^{[r_n]}(.\,|X_n+c_n)$$

respectively, given $X_1, \ldots, X_n, Y_1, \ldots, Y_{3n-3}$. *Then the random variable sequence* $\{X_n\}$ *is called a stochastic-approximation process of type* A_3.

Similarly one can define stochastic-approximation processes of type A_4, A_5, etc.

Now a theorem is stated which is due to Burkholder [10].

THEOREM 4. *Suppose* $\{X_n\}$ *is a type* A_3 *process and* θ *is a real number such that* M *is distribution function with associated density function* f, $H(.\,|X){:}B(.\,|M(X))$ *for*

$$\begin{aligned} B(Y(M(X)) &= 0 && Y < 0,\\ &= 1-M(X) && 0 \leqslant Y < 1 \quad\text{and}\quad 0 \leqslant M(X) < 1,\\ &= 1 && Y \geqslant 1. \end{aligned}$$

If (i) f is increasing for $X < \theta$ and decreasing for $X > 0$.

(ii) If $0 < \delta_1 < \delta_2 < \infty$ then

$$\inf_{\delta_1 \leqslant |X-\theta| \leqslant \delta_2} \left|\frac{f(X-\epsilon)-f(X+\epsilon)}{\epsilon}\right| > 0 \quad \text{for} \quad 0 < \epsilon < \delta_1.$$

(iii) $c_n \to 0$ as $n \to \infty$, $\sum_1^\infty a'_n = \infty$, $\sum_1^\infty \left(\frac{a'_n}{c_n^2}\right)^2 < \infty$.

then
$$P[\lim_{n\to\infty} X_n = \theta] = 1.$$

Proof. Since the result can be proved on the same lines as Theorem 4*c* of Chapter 2, the proof is omitted.

In the next section we will discuss how to make use of this method in sequential regression analysis. Thus illustrative example will show the utility of the method of Burkholder [10].

5. Illustrative example

Now we apply the method discussed in the previous section to a simple regression problem.

Let $Y_j = \alpha - \beta(X_j - \theta)^3 + e_j, j = 1, 2, \ldots, n$ be a regression model with Y_j dependent and X_j independent variable and
$$E(Y_j) = M(X_j) \quad \text{and} \quad M(\theta) = \alpha.$$
The e_j's are independent random variables having common distribution independent of X_j with $E(e_j) = 0$ and $E(e_j^2) = \sigma^2$; where α, β and σ^2 are known real numbers. The problem is to find sequentially an estimate of θ. We use a method of stochastic process of type A_3 to determine an estimate of θ. The mean square error for a finite number of observations is considered. This enables us to make a probability statement about an estimate by Chebyshev's inequality. For computation, the recursion relation is indicated. It is shown under suitable conditions that the mean square error tends to zero as the number of observations tends to infinity.

Example.

(i) Let $Y_n = \alpha - \beta(X_n - \theta)^3 + e_n$ for a natural number n with X_n being independent variable.

(ii) α and β are known real numbers and $\beta > 0$.

(iii) e_n's are independent and have common distribution, independently of the X_n. $E(e_n) = 0$ and $E(e_n^2) = \sigma^2$ (known).

(iv) Let each of $\{a_n\}$ and $\{c_n\}$ be a positive number sequence and
$$Y_{3n-2} = \alpha - \beta(X_n - c_n - \theta)^3 + e_{3n-2},$$
$$Y_{3n-1} = \alpha - \beta(X_n - \theta)^3 + e_{3n-1},$$
$$Y_{3n} = \alpha - \beta(X_n + c_n - \theta)^3 + e_{3n}.$$

(v) Let X_1 be a random variable. Let

$$X_{n+1} = X_n - \frac{a_n}{c_n^2}\left[Y_{3n-1} - \frac{Y_{3n-2}+Y_{3n}}{2}\right].$$

Then

$$\left.\begin{aligned} E(X_{n-1}-\theta)^2 &= E(X_1-\theta)^2 \prod_{k=1}^{n}(1-3a_k\beta)^2 \\ &\quad + \frac{3\sigma^2}{2}\sum_{k=1}^{n}\frac{a_k^2}{c_k^4}\prod_{v=k+1}^{n}(1-3a_v\beta)^2, \\ E(X_{n+1}-\theta)^2 &= E(X_n-\theta)^2 - \frac{2a_n}{c_n^2}E\left\{(X_n-\theta)\left[Y_{3n-1}\right.\right. \\ &\quad \left.\left. - \frac{Y_{3n-2}+Y_{3n}}{2}\right]\right\} \\ &\quad + \frac{a_n^2}{c_n^4}E\left[Y_{3n-1} - \frac{Y_{3n-2}+Y_{3n}}{2}\right]^2. \end{aligned}\right\} \quad (1)$$

The conditional expectation of

$$\left[Y_{3n-1} - \frac{Y_{3n-2}+Y_{3n}}{2}\right]$$

given X_n is

$$3c_n^2\beta(X_n-\theta) \quad (2)$$

and

$$E\left[Y_{3n-1} - \frac{Y_{3n-1}+Y_{3n}}{2}\right]^2 = \frac{3\sigma^2}{2} + 9c_n^4\beta^2 E(X_n-\theta)^2. \quad (3)$$

Substituting (2) and (3) in (1) gives

$$E(X_{n+1}-\theta)^2 = E(X_n-\theta)^2(1-3a_n\beta)^2 + \frac{3\sigma^2}{2}\frac{a_n^2}{c_n^4}. \quad (4)$$

Using this recurrence relation we obtain

$$E(X_{n+1}-\theta)^2 = E(X_1-\theta)^2\prod_{k=1}^{n}(1-3a_k\beta)^2 + \frac{3\sigma^2}{2}\sum_{1}^{n}\frac{a_k^2}{c_k^4}\prod_{v=k+1}^{n}(1-3a_v\beta)^2. \quad (5)$$

The error consists of two parts, the first part being error due to choice and the second part that due to sampling.

Recursion relation.

Let us consider the case when $a_k = c/k$ and $c_k = (c/k)^\delta$, where c is a positive constant and k a positive integer and $\delta > 0$. Then

$$E(X_{n+1}-\theta)^2 = E(X_1-\theta)^2 \prod_{k=1}^{n}\left(1-\frac{3c\beta}{k}\right)^2 + \frac{3(3\beta)^{4\delta-2}\sigma^2}{2} \times \sum_{k=1}^{n}\left(\frac{3c\beta}{k}\right)^{2-4\delta} \prod_{v=k+1}^{n}\left(1-\frac{3c\beta}{v}\right)^2. \quad (6)$$

Let
$$u_n(3c\beta) = \prod_{k=1}^{n}\left(1-\frac{3c\beta}{k}\right)^2, \quad (7)$$

$$\phi_n(\delta, 3c\beta) = \sum_{k=1}^{n}\left(\frac{3c\beta}{k}\right)^{2-4\delta} \prod_{v=k+1}^{n}\left(1-\frac{3c\beta}{v}\right)^2. \quad (8)$$

The recursion relation

$$\phi_n(\delta, c) = \left(\frac{c}{n}\right)^{2-4\delta} + \left(1-\frac{c}{n}\right)^2 \phi_{n-1}(\delta, c) \quad (9)$$

can be used to compute (8).

Asymptotic consideration of mean square error (6).

Let us assume that $0 < \delta < \frac{1}{4}$ and $0 < 3c\beta < 1$

$$\begin{aligned} u_n(3c\beta) &= \prod_{k=1}^{n}\left(1-\frac{3c\beta}{v}\right)^2 \\ &= \exp\left\{2\sum_{k=1}^{n}\log\left(1-\frac{3c\beta}{k}\right)\right\} \\ &\leqslant \exp\left\{-6c\beta\sum_{1}^{n}\frac{1}{k}\right\} \quad \text{since } \log x \leqslant x-1 \\ &\leqslant D_1 n^{-6c\beta} \quad \text{because } \sum_{1}^{n}\frac{1}{k} \geqslant \log(n+1)+1-\log 2. \end{aligned} \quad (10)$$

Let D_1, D_2 and D_3, be appropriate constants.

$$\begin{aligned} \phi_n(\delta, 3c\beta) &= \sum_{k=1}^{n}\left(\frac{3c\beta}{k}\right)^{2-4\delta} \prod_{v=k+1}^{n}\left(1-\frac{3c\beta}{v}\right)^2 \\ &= (3c\beta)^{2-4\delta}\sum_{k=1}^{n} k^{4\delta-2}\frac{\prod_{v=1}^{n}\left(1-\frac{3c\beta}{v}\right)^2}{\prod_{v=1}^{k}\left(1-\frac{3c\beta}{v}\right)^2} \end{aligned}$$

$$\leqslant (3c\beta)^{2-4\delta}(D_1 n^{-6c\beta}) \sum_{k=1}^{n} \frac{k^{4\delta-2}}{\prod_{v=1}^{k}\left(1-\frac{3c\beta}{v}\right)^2}.$$

$$\frac{1}{\prod_{v=1}^{k}\left(1-\frac{3c\beta}{v}\right)^2} \leqslant D_2 k^{6c\beta},$$

since $\quad 1-\frac{1}{x} \leqslant \log x \quad$ and $\quad \sum_{v=2}^{k} \frac{1}{v-1} \leqslant 1+\log(k-1).$

Therefore

$$\phi_n(\delta, 3c\beta) \leqslant (3c\beta)^{2-4\delta}(D_1 n^{-3c\beta})\, D_2 \sum_{k=1}^{n} k^{4\delta-2+6c\beta}$$
$$\leqslant D_3 n^{4\delta-1}.$$

Since $\quad 0 < \delta < 1/4, \quad \phi_n(\delta, 3c\beta) \to 0 \quad$ as $\quad n \to \infty. \qquad (11)$

Thus from (10) and (11) it follows that $\lim_{n\to\infty} E(X_{n+1}-\theta)^2 = 0.$

REMARK. One can prove this result very easily by using Kronecker's lemma, but the technique employed for actual computation will be useful.

6. Problems

1. Assume the conditions of Theorem 2 and let

$$|M'''(x)| \leqslant Q \quad \text{for some} \quad Q > 0.$$

If (I), (II), (10), (11) and (12) hold then prove that

$$b_n = \begin{cases} O\left(\frac{1}{n^{4r}}\right) & \text{when} \quad r < \frac{\alpha}{6}, \\ O\left(\frac{1}{n^{\alpha-2r}}\right) & \text{when} \quad r \geqslant \frac{\alpha}{6}. \end{cases}$$

2. Assume the conditions of Theorem 2. Let

$$M^{(2k+1)}(\theta) = 0, \quad |M^{(2k+2)}(x)| \leqslant A^{2k+1}(2k+1)! \quad (k = 1, 2, \ldots),$$

for every x and some $A > 0$. If (I) (II), (10), (11) and (12) hold then prove that

$$b_n = O\left(\frac{1}{n^{\alpha-2r}}\right). \qquad \text{(Dupač)}$$

3. Let $Y(x)$ be a scalar valued random variable with distribution function $H(y|x)$, where x is a scalar valued parameter. Let $M(x) = \int y dH(y|x)$ and $M(x)$ be continuous and have a unique local minimum $x = 0$ say $\theta = 0$ and let $\{a_n\}, \{c_n\}$ be sequences of positive real numbers

$$\sum_1^\infty a_n = \infty, \quad \sum_1^\infty a_n^2 c_n^{-2} < \infty, \quad \sum_1^\infty a_n c_n < \infty \quad (c_n \to 0 \text{ as } n \to \infty).$$

$$X_{n+1} = X_n + a_n M_{c_n}(X_n) + a_n \xi_n / c_n,$$

$$M_c(X) = [M(X+c) - M(X-c)]/c,$$

$$\xi_n = [Y(X_n + c_n) - M(X_n + c_n)] - [Y(X_n - c) - M(X_n - c_n)].$$

Let K_0, K and C_0 be positive real numbers. Let

$$E[\xi_n | X_0, X_1, \ldots, X_n] = O, \quad E[\xi_n^2 | X_0, X_1, \ldots, X_n] \leqslant 2\sigma^2 < \infty$$

with probability one. For $0 < c < C_0 < \infty$, let

$$-cKX^2 \leqslant [M(X+c) - M(X-c)]X \leqslant -K_0 X^2,$$

$$a_n = A/n^{1-\epsilon}, \quad c_n = c/n^{\frac{1}{2}-\eta} < C_0, \quad \eta > \epsilon > 0, \quad \eta + \epsilon > \tfrac{1}{2}.$$

For each integer $r > 0$, there is a positive number $M_r < \infty$ such that

$$E[|Y(X_n) - M(X_n)|^r | X_0, \ldots, X_n|] \leqslant M_r/2 \qquad (KA \leqslant 1 \text{ and } A > 0).$$

Then prove that for all integral m, N such that $\infty > m \geqslant N \geqslant 1$ and even integer r

$$P[\max_{m \geqslant n \geqslant N} |X_n| > \epsilon] < (EX_N^r + \delta_{Nr})/\epsilon^r,$$

where δ_{Nr} is finite and tends to zero as $A \to 0$ and also prove that $X_n \to 0$ as $n \to \infty$ with probability one. (Kushner)

4. Let $\{X_n\}$ be stochastic-approximation process type A_2, such that $\Sigma a_n = \infty$ and

$$\Sigma \left(\frac{a_n}{c_n}\right)^2 < \infty$$

and with recurrence relation

$$X_{n+1} = X_n - a_n B_n^{-1} \left[\frac{Y''_n - Y'_n}{2C_n}\right]$$

and B_n is defined as follows: Let $\{Y'''\}$ be a sequence of random variables such that Y'''_n is, conditional on given

$$\{Y'_k, Y''_k, Y'''_k, \quad 1 \leqslant k \leqslant n-1\},$$

independent of $Y'_n - Y''_n$ and distributed according to the distribution of $Y(X_n)$. Then, let

$$B^* = n^{-1} \sum_{k=1}^{n} [-(Y'_n - 2Y''_n + Y'''_n)/c_n^2]$$

and
$$B_n = \begin{cases} b' & \text{if } B_n^* < b', \\ B_n^* & \text{otherwise}, \\ b'' & \text{if } B_n^* > b'', \end{cases}$$

where b' and b'' are such that $0 < b' < b'' < \infty$,

Furthermore the following conditions are satisfied.

(i) $M(x)$ is Borel measurable.

(ii) $M(x)$ has unique maximum

(iii) For every $\epsilon > 0$, there exists c_ϵ such that $0 < c_\epsilon < \epsilon$ and

$$\inf_{\substack{\epsilon \leqslant |x-\theta| \leqslant 1/\epsilon \\ 0<c<c_\epsilon}} (x-\theta) \frac{M(x-c) - M(x+c)}{2c} > 0.$$

(iv) There exists a constant c_1 such that for all c with $0 < c < c_1$ and for all $x \in R$,

$$|M(x+c) - M(x)| < k_1 + k_2|x|, \quad \text{where} \quad k_1 \quad \text{and} \quad k_2 > 0.$$

(v) There exists a positive constant c_2 such that

$$\sup_{x\in R} \sup_{0<c<c_2} |M(x+c) - 2M(x) + M(x-c)| < \infty.$$

(vi) $\sup_{x\in R} \operatorname{var} Y < \infty$.

(a) $na_n = 1 + O(n^{-\frac{1}{2}})$ as $n \to \infty$,

(b) $c_n \to cn^{-\gamma}$ as $n \to \infty$ with $0 < \gamma < \frac{1}{2}$.

Prove that $X_n \to \theta$ with probability one and $E(X_n - \theta)^2 \to 0$ as $n \to \infty$. (Venter)

5. Let $\{X_n\}$ be stochastic-approximation process of type A_2 with following assumptions:

(i) $M(x)$ is Borel measurable.

(ii) $M(x)$ has unique maximum.

(iii) Assume condition (iii) of problem 4.

(iv) There exists a constant c_1 such that for all c with $0 < c < c_1$ and for all $x \in R$, $|M(x+c)-M(x)| < k_1+k_2|x|$, where $k_1, k_2 > 0$.

(v) For every $x \in R$, $M(x) = \beta_0 - \beta/2(x-\theta)^2 + \delta(x)$, where $\beta_0 \in R$ and β is a positive constant, while δ is such that

$$\delta(x) = O(|x-\theta|^{2+\rho}) \quad \text{as} \quad |x-\theta| \to 0, \quad \text{where} \quad \rho > 0.$$

(vi) There exist positive constants ϵ_0, c_0 and k_0 such that $\epsilon_0 > c_0$ and such that for all $c < c_0$ and for all x satisfying

$$c < |x-\theta| < \epsilon_0,$$

we have
$$(x-\theta)\frac{M(x-c)-M(x+c)}{2c} > k_0(x-\theta)^2.$$

There exists $c_2 > 0$, such that

$$\sup_{x \in R} \sup_{0<c<c_2} |M(x+c)-2M(x)+M(x-c)| < \infty,$$

(vii) $\sup_{x\in R} \operatorname{var} Y < \infty$.

(*a*) β and $k_0 \geqslant \min[\frac{1}{2}-\gamma, (1+\rho)\gamma]$.

(*b*) $na_n \to 1$ as $n \to \infty$.

(*c*) $c_n = O(n^{-\gamma})$ as $n \to \infty$, $0 < \gamma < \frac{1}{2}$.

(*d*) $\lambda < \min[\frac{1}{2}-\gamma, (1+\rho)\gamma]$.

Prove that

$$X_n - \theta = O(n^{-\lambda}) \text{ with probability one as } n \to \infty.$$

Hint.
$$\begin{aligned}(n+1)^\lambda X_{n+1} = n^\lambda X_n - n^{-1}\Big[&(1+\epsilon_{1n}) I_{[|X_n|>c_n]} \times \frac{M(X_n-c_n)-M(X_n+c_n)}{2c_n X_n} \\ &+\beta(1+\epsilon_{2n}) I_{[|X_n|\leqslant c_n]} - \lambda(1+\epsilon_{3n})\Big] n^\lambda X_n \\ &- n^{\lambda-1}(1+\epsilon_{4n}) I_{[|X_n|\leqslant c_n]} \frac{\delta(X_n-c_n)-\delta(X_n+c_n)}{2c_n} \\ &- n^{\lambda-1}(1+\epsilon_{5n})\frac{Z'_n - Z''_n}{2c_n},\end{aligned}$$

where
$$Z'_n = Y'_n - M(X_n+c_n),$$
$$Z''_n = Y''_n = M(X_n-c_n)$$

and $\epsilon_{1n}, \epsilon_{2n}, \ldots, \epsilon_{5n} \to 0$ as $n \to \infty$ and these are non-random variables. (Venter)

CHAPTER 4

APPLICATIONS

1. Introduction

In Chapter 2 we discussed various mathematical models of the Robbins–Monro method and in this chapter we take up problems of their utility. Stochastic-approximation procedures require very little prior knowledge of the process (of its input, etc.) and achieve reasonably good results.

First we discuss the application of stochastic approximation to adaptive control processes. This can be viewed as an extension of the conventional proportional control procedure discussed by Box and Jenkins [7].

Secondly we look at a problem of pharmacology and show how one can make use of these methods with advantage. Thirdly an application to a problem of reliability is sought. And finally we discuss a method to estimate sequentially quantal response by the Robbins–Monro method.

It is interesting to see that these methods work satisfactorily in practice. Wetherill [81] used simulation methods in studying the application of stochastic-approximation procedures and suggested several modifications.

2. Adaptive control processes

(*a*) *A process control problem.*

The objective of process control may be to hold a response Y as close as possible to some specified target Y_0 by adjusting an independent variable X. Let us look at this problem in the context of a simple chemical process. Supoose that our objective is to hold the viscosity at the output of a process as near to Y_0 as possible. The only control we plan to use is a value controlling cooling water to a heat exchanger in the process. The value can be turned from a fully closed setting which we shall call X_L to a fully open setting which we shall call X_u. Initially we set the valve at some intermediate position X_1. Exactly 15 minutes later we observe the viscosity Y_1 at the output and use the observed error $E_1 = Y_1 - Y_0$ to help us readjust the valve setting to a new value to be held for the next 15 minute interval. Thus we sequentially adjust the value to a new setting (X_{n+1}) every 15 minutes on the basis of our knowledge of all prior errors ($E_1, \ldots, E_n$) and all prior valve settings ($X_1, \ldots, X_n$). In this section it is assumed that an increase in the setting of the valve (X) is known to tend to cause an increase in the true viscosity, that is, we know the direction in which to turn the valve to compensate for excessively high or low viscosity. This type of problem is discussed by Comer [16] whence this material has, more or less, been taken.

The primary objective in choosing parameters for any proposed control procedure will be to minimize the true asymptotic loss (L_T) defined as the asymptotic value of the mean square deviation of true viscosity from Y_0. This asymptotic loss will always be assumed to exist for models studied in this section. Any measuring errors that may exist are assumed to form a stationary sequence of independent random variables with variance σ^2, and the stochastic behaviour of this sequence is assumed to be independent of the random variation associated with the process. Under these assumptions the true asymptotic loss will be minimized if we can minimize

$$L = \lim_{n\to\infty} E(Y_n - Y_0)^2 = \sigma^2 + L_T, \tag{1}$$

where L is now based on the observed values of Y_n and will be referred to as the asymptotic loss.

A secondary objective in choosing parameters for any control procedure will be to keep the loss small at each iteration point n. We shall define this loss as

$$L_n = E(Y_n - Y_0)^2. \tag{2}$$

We shall now discuss a simple proportional control procedure and show how one can meet the primary and secondary objectives of process control.

(*b*) *Proportional control procedures.*

The simple proportional control procedure requires that we choose a control parameter a and make adjustments to the valve setting sequentially on the basis of the recursive calculation

$$X_{n+1} = X_n - a(Y_n - Y_0) \quad (n = 1, 2, \ldots). \tag{3}$$

In this procedure X_1 and a are chosen on the basis of the past experience. In practice the range of X_{n+1} is limited to the interval $[X_L, X_u]$. For example, should any calculated value of X_{n+1} lie below X_L, X_{n+1} is chosen equal to X_L.

Now we give an example where additional information is available on Y's and we are interested in controlling the process at the inflection point which is assumed to exist and discuss how we can meet the secondary object of process control.

(*c*) *Examples.*

Let

(i) $c_n = c$ and $\dfrac{a_n}{c_n^2} = a$ for every n.

(ii) $Y_n(X)$ be stationary random variable in wide sense, with covariance function

$$B(i) = E[(Y_n - E(Y_n))\,(Y_{n+i} - E(Y_{n+i}))].$$

(iii) $Y_n(X_n - c)$, $Y_n(X_n)$ and $Y_n(X_n + c)$ be independent for every n.

(iv) $Y_n(X_n) = \alpha - \beta(X_n - \theta)^3 + \epsilon_n$, where α and β are known constants and $\beta > 0$ and $\operatorname{var}[Y_n(X_n)] = \sigma^2$ (known).

(v) $\psi_n = \left[Y_{3n-1} - \dfrac{Y_{3n-2} + Y_{3n}}{2}\right]$.

(vi) $z_n = \psi_n - 3c^2\beta(X_n - \theta)$.

(vii) $\rho = 1 - 3a\beta c^2$.

Then

$$\Delta = \lim_{n\to\infty} E(\psi_n^2) = \frac{3}{1+\rho}\left[\sigma^2 - (1-\rho)\sum_1^\infty \rho^{i-1}B(i)\right].$$

Proof. Let

$$\begin{aligned} X_{n+1} - \theta &= (X_n - \theta) - az_n - \frac{3\beta ac^2}{2}(X_n - \theta) \\ &= (1 - 3\beta ac^2)\,(X_n - \theta) - az_n, \end{aligned}$$

$$(X_n-\theta) = -a[\rho^{n-2}z_1+\rho^{n-3}z_2+\ldots+z_{n-1}],$$

$$\psi_n = z_n - 3ac^2\beta[z_{n-1}+\rho z_{n-2}+\ldots+\rho^{n-2}z_1].$$

$$\psi_n = z_n-(1-\rho)[z_{n-1}+\rho z_{n-2}+\ldots+\rho^{n-2}z_1],$$

$$E(\psi_n^2) = E\left[z_n-(1-\rho)\sum_1^{n-1}\rho^{i-1}z_{n-i}\right]^2,$$

$$\Delta = \lim_{n\to\infty} E(\psi_n^2) = \tfrac{3}{2}\sigma^2\left[1+\frac{(1-\rho)^2}{1-\rho^2}\right]-\tfrac{3}{2}(1-\rho)\sum_1^{\infty}\rho^{i-1}B(i)$$

$$+\tfrac{3}{2}(1-\rho)^2\frac{2\rho}{1-\rho^2}\sum_1^{\infty}\rho^{i-1}B(i)$$

$$= \frac{3}{1+\rho}\left[\sigma^2-(1-\rho)\sum_1^{\infty}\rho^{i-1}B(\mathrm{i})\right].$$

REMARK. In control processes one is interested in running the process at a fixed level (say at zero), in which case the process variable is ψ_n. It is desirable to consider $E(\psi_n^2)$ and determine a's for given c which minimizes Δ.

Example (*a*). Let the covariance function of a process be $B(i) = e^{-\lambda i}\sigma^2$, where $\lambda > 0$ and i positive integer. Then

$$\Delta = \frac{3\sigma^2}{1+\rho}\left[1-(1-\rho)\sum_1^{\infty}\rho^{i-1}e^{-\lambda i}\right]$$

$$= \frac{3\sigma^2}{1+\rho}\left[1-\frac{(1-\rho)e^{-\lambda}}{1-\rho e^{-\lambda}}\right] = \frac{3\sigma^2(1-e^{-\lambda})}{(1+\rho)(1-\rho e^{-\lambda})}.$$

Then Δ attains minimum when

$$a = \frac{1}{\beta c^2}-\frac{1}{3e^{-\lambda}\beta c^2}.$$

Hence in this case one can determine the optimum value of a for given c and λ.

Example (*b*). Let the covariance function of a process be

$$B(i) = \eta e^{-\lambda i}\cos(\tau i)\,\sigma^2,$$

$$\Delta = \frac{3\sigma^2}{1+\rho}\left[1-(1-\rho)\sum^{\infty}\eta\rho^{i-1}e^{-\lambda i}\cos\tau i\right]$$

$$= \frac{3\sigma^2}{1+\rho}\left[1-\frac{1-\rho}{\rho}\eta\sum_1^{\infty}(\rho e^{-\lambda})^i\cos\tau i\right]$$

$$= \frac{3\sigma^2}{1+\rho}\left[1-\frac{1-\rho}{\rho}\eta\frac{1-\rho\cos\tau e^{-\lambda}}{1-2\rho e^{-\lambda}\cos\tau+\rho^2e^{-2\lambda}}\right].$$

For a given value of τ one can determine value of a which minimizes Δ.

An interesting generalization of the simple proportional procedure arises when we allow the parameter a to vary with time. The recursive calculation then becomes

$$X_{n+1} = X_n - a_n(Y_n - Y_0). \tag{4}$$

An intuitively appealing special case occurs when $\{a_n\}$ consists of positive numbers converging to zero. We might expect to be able to apply such a procedure to help us choose a good valve setting when a process like the one described in the previous section is started up or after it has been subjected to some disturbance. Our hope is that the relatively large early adjustments in X_n will permit X_n to move fairly rapidly toward a good value around which it will hover as adjustments become smaller. We shall give such a stochastic-approximation procedure after discussing the general process although such procedures have already been investigated theoretically in Chapters 2 and 3.

(*d*) *The general process.*

In order to appraise the performance of various control methods, it will be necessary to set up a mathematical model of the process. This model must show the relationship between Y_n and X_n and must also include the effect of undesirable random variation on Y_n. To develop the general model of the process used, we start by defining the steady-state effect of an adjustment in X to be

$$Y_n^{(1)} = M(X_n), \tag{5}$$

where $M(X)$ is a strictly increasing regression function, and for all distinct X' lying in the specified interval $[X_L, X_u]$ there exist bounds β_L and β_u such that

$$0 < \beta_L \leqslant \{M(X) - M(X')\}/(X - X') \leqslant \beta_u < \infty. \tag{6}$$

We also assume that there exist a value θ in the interval $[X_L, X_u]$ such that $M(\theta) = Y_0$, where Y_0 is the desired value of the response.

When X is changed frequently, the effect of the adjustment may be only partially realized in one time interval. The effect of an adjustment in X under these conditions may include a first-order dynamic lag and is then given by

$$Y_n^{(1)} = Y_{n-1}^{(1)} + (1-k)\{M(X_n) - Y_{n-1}^{(1)}\}, \tag{7}$$

where k represents the first-order dynamic lag in the process and has values such that $0 \leqslant k < 1$. When no lag exists, we take the value k to be zero and (7) reduces to (5).

Finally, we can define Y_n as

$$Y_n = Y_n^{(1)} + Z_n, \tag{8}$$

where $\{Z_n\}$ is a sequence of stationary random variables with $E(Z_0) = 0$ and finite $E(Z_n^2)$, whose stochastic behaviour is independent of the control procedure used, and whose correlation structure satisfies the following condition. There exists a sequence of non-negative numbers $\{\xi_n\}$ such that

$$\lim_{n\to\infty} \xi_n = 0 \tag{9}$$

and
$$[E\{E(Z_n | Z_{n-m}, Z_{n-m-1}, \ldots, Z_1)\}^2]^{\frac{1}{2}} \leqslant \xi_m \tag{10}$$

for all $n = 1, 2, \ldots$ and for all $m \leqslant n$. By combining (7) and (8) we find that Y_n can be expressed in terms of X_n and Z_n as

$$Y_n = (1-k)\,M(X_n) + Z_n + kY_{n-1} - kZ_{n-1}. \tag{11}$$

Note that when lag is negligible ($k = 0$), (11) reduces to

$$Y_n = M(X_n) + Z_n. \tag{12}$$

Such case of stochastic approximation is discussed in Theorem 4*c* of Chapter 2.

(*e*) *Proposed control procedure for choosing a good fixed value of X.*

In the context of chemical processes and the preceding discussion we can propose the following procedure for which it is very easy to establish mathematical convergence in mean square and with probability one to the proposed value of the generated sequence by means of the results of Chapter 2.

(i) Let X_1 in the permissible range $[X_L, X_u]$ be the best *a priori* estimate of θ.

(ii) Define a series of constants $\{a_n\}$ such that

$$a_n > 0, \quad \lim_{n\to\infty} a_n = 0, \quad \sum_1^\infty a_n = -\infty. \tag{13}$$

(iii) Recursively define X_{n+1} for $n = 1, 2, \ldots$ as

$$X_{n+1} = \begin{cases} X_L & [\{X_n - a_n(Y_n - Y_0)\} < X_L], \\ X_n - a_n(Y_n - Y_0) & [X_L < X_n - a_n(Y_n - Y_0) \leqslant X_u], \\ X_u & [\{X_n - a_n(Y_n - Y_0)\} > X_u]. \end{cases} \tag{14}$$

The performance of this procedure for some specific choices of X_1 and the sequence $\{a_n\}$ is described below.

Motivation for the procedure is simple. The direction in which X_n is changed should tend to bring Y_n closer to Y_0. Adjustments to X_n also tend to become smaller as n increases and we have

$$\lim_{n\to\infty} E(X_n - \theta)^2 = 0 \tag{15}$$

provided appropriate restrictions such as those described for a general process are placed on the process. Although it is easy to obtain good asymptotic results using this procedure, practical considerations dictate that we choose the sequence $\{a_n\}$ and the initial setting X_1 to make the performance reasonably good even for small value of n. Choice of X_1 to a value $X_1 - \theta = 1/\beta$ and $a_n = a/n$ give a reasonably good result. Comer has also discussed the case where $a_n \to a > 0$ as $n \to \infty$ and given many illustrative processes.

3. Two regression functions for a kinetic model

In the literature of pharmacology several kinetic models for joint action of drugs are discussed to which stochastic-approximation methods can be applied with advantage In this section, a simple model of competitive antagonism is considered and it is shown how one can make use of the Robbins–Monro method. These problems are discussed in Perrin [62] and Epling [30].

(*a*) *Competitive antagonism.*

Let us assume that a drug, which we shall call the agonist, produces some measurable biological response in an organism by combining directly with a specific group of receptors in that organism and that the response elicited is, within a certain range, an increasing function of the proportion of receptors in combination with the agonist. Further, assume that a second drug, the antagonist, when present may combine with receptors of the same type in a manner such as to inactivate them or make then unavailable for action with the agonist so that if the two drugs are administered jointly the antagonist effectively reduces the fraction of receptors in combination with the agonist and produces an inhibitory effect on the response. This is often thought to be the model for competitive antagonism between drugs. In the following section this problem is formulated mathematically and it is shown how one can make use of the Robbins–Monro method.

(*b*) *Mathematical formulation.*

Let us consider two families of random variables, $\{Y_1(x)\}$ and $\{Y_2(x)\}$ with corresponding distribution, expectation and variance $H_i(Y(x))$, $M_i(x)$, $\sigma_i^2(x)$, $i = 1, 2$. Further suppose that $M_1(x)$ and $M_2(x)$ have the following properties.

(i) Each $M_i(x)$ is a strictly monotone increasing continuous function of x.

(ii) Given any real α, there exists a pair (θ_1, θ_2) such that

$$\theta_1 + \theta_2 = \alpha \quad \text{and} \quad M_1(\theta_1) = M_2(\theta_2) \tag{1}$$

denote the common value of $M_i(\theta_i(\alpha))$ by $M(\alpha)$, and let $\{a_n\}$ a positive number sequence have the following properties,

$$a_n > 0 \quad \text{for all } n, \quad \sum_1^\infty a_n = \infty \quad \text{and} \quad \sum_1^\infty a_n^2 < \infty. \tag{2}$$

Pairs of experiments are to be performed sequentially in the following manner. Let x_1, and x_2, be two arbitrary initial levels, and $Y_1(x)_1$ and $Y_2(x_2)$ be observed values. Perform a pair of experiments at levels X_{12} and X_{22} determined by

$$X_{i2} = X_{i1} + a_1[\overline{Y}_1 - Y_i(x_1)],$$

where $$\overline{Y}_1 = [Y_1(x_{11}) + Y_2(x_{21})]/2.$$

In general, letting x_{1n} and x_{2n} denote the levels for the nth pairs of experiments, $Y_{1n} = Y_1(x_{1n})$ and $Y_{2n} = Y_2(x_{2n})$ denote the observed responses for the nth pair of experiments, and

$$\bar{Y}_n = [Y_{1n}+Y_{2n}]/2.$$

The levels for the $(n+1)$st pair of experiments are

$$X_{in+1} = X_{in}+a_n(\bar{Y}_n-Y_{in}) \quad (i = 1, 2, n \geqslant 1). \tag{3}$$

Assume that for given X_{1n} and X_{2n}, Y_{1n} and Y_{2n} are independent. Let

$$\sigma_i^2(x) < v < \infty \quad \text{for} \quad i = 1, 2, \quad \text{for all } x. \tag{4}$$

There exist positive constants A and B such that

$$|M_i(x)| < A|x|+B < \infty \quad \text{for} \quad i = 1, 2 \quad \text{for all } x. \tag{5}$$

Under these conditions it is shown that sequences $\{X_{1n}\}$ and $\{X_{2n}\}$ defined by (3) converge in mean square and with probability one to θ_1 and θ_2 respectively and $M_1(\lim_{n\to\infty} X_{1n}) = M_2(\lim_{n\to\infty} X_{2n})$ with probability one. Let us suppose that X_{1n} and X_{2n} converge with probability 1 to limits L_1 and L_2 say. By definition,

$$X_{1n+1}+X_{2n+1} = X_{1n}+X_{2n} = \alpha$$

for all n; hence $L_1+L_2 = \alpha$. If $L_1 \neq \theta_1$ say $L_1 < \theta_1$ then $L_2 > \theta_2$ and from the strict monotonicity in the neighbourhood of θ_1, for any $\delta > 0$, there exists n_0 and a positive ϵ such that

$$M_2(X_{2n})-M_1(X_{1n}) > \epsilon$$

for all $n > n_0$ with probability greater than $1-\delta$. But since $\sum_{n=1}^{\infty} a_n = \infty$ this means that

$$X_{1n+1} = X_{1n}+a_n[M_2(X_{2n})-M_1(X_{1n})]/2+a_n\delta_n,$$

where $E(\delta_n) = 0$, does not converge to a real number which is a contradiction. Thus if x_{1n} and x_{2n} do converge, they must converge to θ_1 and θ_2 given by (1).

For each possible $\alpha = X_{11}+X_{21}$, define the family of random variables

$$W_\alpha(r) = [Y_2(\theta_2+r)-Y_1(\theta_1-r)]/2 \quad (-\infty < r < \infty).$$

$E\{W_\alpha(r)\}$ is a strictly monotone increasing function of r, and $E[W_\alpha(0)] = 0$, $\sigma_\alpha^2(r) = [\sigma_2^2(\theta_2+r)+\sigma_1^2(\theta_1-r)]/4$ is the variance of $W_\alpha(r)$.

(c) *One dimensional stochastic approximation.*

Construct a sequence $\{r_n\}$ $(n \geqslant 1)$ in which r_1 is chosen arbitrarily, and the succeeding levels are determined recursively by

$$r_{n+1} = r_n - a_n w_\alpha(r_n), \tag{6}$$

where $w_\alpha(r_n)$ is observed response at the level r_n. For each fixed α, the rule $\{X_{1n} = \theta_1 - r_n,\ X_{2n} = \theta_2 + r_n\}$ establishes a 1-1 correspondence between possible sequences $\{(X_{1n}, X_{2n})\}$ $(n \geqslant 1)$ as defined before and sequence $\{r_n\}$ as defined here. The convergence of r_n to 0 is equivalent to the convergence of X_{1n} to θ_1 and X_{2n} to θ_2.

The random variable family $\{w_\alpha(r)\}$ obviously satisfies all conditions of Dvoretzky's theorem except perhaps condition (4) which seems to be true also and it can be checked as follows.

$$\begin{aligned}|E\{w_\alpha(r)\}| &= \tfrac{1}{2}|M_2(\theta_2+r) - M_1(\theta_1-r)| \\ &\leqslant \tfrac{1}{2}[|M_2(\theta_2+r)| + |M_1(\theta_1-r)|] \\ &\leqslant \tfrac{1}{2}[A(|\theta_2+r| + |\theta_1-r|) + 2B] \\ &\leqslant A|r| + [A(|\theta_1| + |\theta_2|) + B] < \infty.\end{aligned}$$

Since $E\{W_\alpha(0)\} = 0$, the Robbins–Monro sequence $\{r_n\}$ defined by (6) converges to 0 in mean square and with probability one. Thus, whatever the initial levels, there exist θ_1 and θ_2 (depending on these initial levels) such that X_{in} converges to θ_i in mean square and with probability one. And $M_1(\lim_{n\to\infty} X_{1n}) = M_2(\lim_{n\to\infty} X_{2n})$ holds with probability one by the assumption (1).

Thus the two-curve problem is equivalent to a one-dimensional difference process whose convergence to zero implies the desired result.

Similarly one can consider more than two drugs in the kinetic model and determine the effective dose levels as has been done in the case of θ_1 and θ_2 in the preceding discussion.

4. Application of stochastic approximation to a problem of reliability

In the following problem we consider a 'system' or 'item', with a 'life time' that has the distribution function $F(t)$, and which is inspected at times $t_1, t_2, \ldots$. If inspection reveals that the system is inoperative, it is repaired (or replaced); otherwise nothing is done. The general problem is to choose the inspection plan, i.e. the sequence $t_1, t_2, \ldots$ in an optimal way in a suitable sense. A criterion of optimality is defined and it is proved that a stochastic-approximation plan satisfies the criterion. This problem is discussed in Venter and Gastwirth [71].

Let the system have an exponential lifetime, i.e.

$$F(t) = \begin{cases} 0 & \text{if} \quad t < 0, \\ 1 - e^{-\lambda t} & \text{if} \quad t \geqslant 0 \end{cases}$$

with $\lambda > 0$ which is an unknown parameter. It is assumed that there exist known constants $\boldsymbol{\lambda}$ and $\bar{\lambda}$ with $0 < \boldsymbol{\lambda} < \bar{\lambda}$ such that

$$\lambda \in (\boldsymbol{\lambda}, \bar{\lambda}). \tag{1}$$

Let us specify the inter-inspection times by

$$T_1 = t_1, \quad T_1 = t_i - t_{i-1} \quad (i = 2, 3, \ldots). \tag{2}$$

Let $\{u_n\}$ be an arbitrary sequence of random variables the joint distribution of any finite number of which does not depend on λ. Take $T_1 = \max\{0, u_1\}$ and define $\{T_n\}$ iteratively by

$$T_{n+1} = \max\{0, f_n(Y_1, \ldots, Y_n) + u_n\} \tag{3}$$

for $n = 1, 2, \ldots$. Here, each Y_i, $i = 1, 2, \ldots$ is a random variable with conditional distribution given $\{Y_1, \ldots, Y_{i-1}, T_1, \ldots, T_i\}$ specified by

$$Y_i = \begin{cases} 1 & \text{with probability} \quad e^{-\lambda T_i}, \\ 0 & \text{with probability} \quad 1 - e^{-\lambda T_i}, \end{cases} \tag{4}$$

that is $Y_i = 0$ if the ith inspection reveals that the system is inoperative and $Y_i = 1$ otherwise. Also, f_n is a real-valued measurable function of $(Y_1, \ldots, Y_n)$, functionally independent of λ. Intuitively, after n inspections, the next inspection time T_{n+1}

depends on the past observations $(Y_1, \ldots, Y_n)$ through f_n while u_n allows for additional randomization. The class of all these inspection plans is denoted by g and a generic element of g by I.

(*a*) *Maximization of information.*

The average information obtainable from a plan I after n inspections is defined by

$$J_n(I, \lambda) = n^{-1} E\left[\frac{d}{d\lambda} \log L_n(\lambda)\right]^2, \tag{5}$$

where $L_n(\lambda)$ is the likelihood function of λ based on

$$(Y_1, Y_2, \ldots, Y_n, T_1, \ldots, T_n).$$

Also let

$$J(I, \lambda) = \liminf_{n \to \infty} J_n(I, \lambda), \tag{6}$$

$J(I, \lambda)$ is called the limiting average information obtainable from plan I. The problem is that of maximizing $J_n(I, \lambda)$ and $J(I, \lambda)$ by a judicious choice of I. This method of efficient estimation of λ is well known in the literature on Reliability.

THEOREM. *For each n, for all λ and for all I*

$$J_n(I, \lambda) \leqslant \lambda^{-1} T_\lambda (2 - \lambda T_\lambda), \tag{7}$$

where T_λ is the solution of the equation

$$e^{-\lambda T} = 1 - \tfrac{1}{2}\lambda T. \tag{8}$$

REMARK. $T_\lambda = -\lambda^{-1} \log p$ and T_λ is the 100 $(1-p)$th percentile of exponential distribution where

$$p \approx 0{\cdot}203. \tag{9}$$

Proof. The conditional probability of observing $Y_1, \ldots, Y_n$ given $(u_1, \ldots, u_n)$ is

$$\prod_{i=1}^{n} (e^{-\lambda T_i})^{Y_i} (1 - e^{-\lambda T_i})^{1 - Y_i}. \tag{10}$$

The distribution of $(u_1, \ldots, u_n)$ being independent of λ, we find

$$\begin{aligned} \frac{d}{d\lambda} \log L_n(\lambda) &= -\sum_{i=1}^{n} T_i Y_i + \sum_{i=1}^{n} (1 - Y_i)\, T_i e^{-\lambda T_i} / (1 - e^{-\lambda T_i}) \\ &= -\sum_{1}^{n} T_i (Y_i - e^{-\lambda T_i}) / (1 - e^{-\lambda T_i}) \end{aligned}$$

writing $X_i = T_i(Y_i - e^{-\lambda T_i})/(1-e^{-\lambda T_i})$

one has, for $j > i$

$$E(X_i X_j) = E\{X_i T_j(1-e^{-\lambda T_j})^{-1} \times E[(Y_j - e^{-\lambda T_j})|Y_1, \ldots, Y_{j-1}, T_1, \ldots, T_j]\} = 0$$

because of (4). Hence

$$\begin{aligned} E\left[\frac{d}{d\lambda}\log L_n(\lambda)\right]^2 &= \sum_1^n E(X_i^2) \\ &= \sum_1^n E\{T_i^2(1-e^{-\lambda T_i})^{-2} E[(Y_i - e^{-\lambda T_i})^2 \\ &\qquad \times |Y_1, \ldots, Y_{i-1}, T_1, \ldots, T_i]\} \\ &= E\sum_1^n T_i^2(1-e^{-\lambda T_i})^{-1} e^{-\lambda T_i} \\ &\leqslant nT_\lambda^2(1-e^{-\lambda T_\lambda})^{-1} e^{-\lambda T_\lambda}, \end{aligned}$$

since the function $T^2(1-e^{-\lambda T})^{-1} e^{-\lambda T}$ is maximized by $T = T_\lambda$. Hence the result follows.

Equality in (7) is attained if and only if $T_i = T_\lambda$ with probability one for each i, i.e. if λ were known, the optimal inspection plan in the sense of maximzing $J_n(I, \lambda)$ for each n and λ would call for periodic inspections with inter-inspection times T_λ. However, within class g (i.e. when λ is unknown) there exists no optimal plan giving equality in (7).

Now an optimality criterion is defined which takes into account the information about λ that comes available as inspection proceeds.

DEFINITION. *An inspection plan I is said to be adaptive (relative to $J(I,\lambda)$) if*

$$J(I, \lambda) = \lambda^{-1}T_\lambda(2-\lambda T_\lambda). \tag{11}$$

Now a stochastic-approximation plan is defined and shown to be is adaptive.

(b) A stochastic-approximation plan.

The following plan (denoted by SA) is based on the Robbins-Monro method, and exploits the fact that T_λ corresponds to the 100 $(1-p) \approx 79{\cdot}7$th percentile of the exponential distribution, independent of λ.

DEFINITION. *Choose arbitrary* T_1 *in* $[T_{\bar{\lambda}}, T_{\lambda}\}$ *and after defining* $T_1, \ldots, T_n$ *let*

$$\left.\begin{aligned} \lambda_n &= T_n^{-1}\log p, \\ A_n = \lambda_n^{-1}p^{-1} &= -T_n(p\log p)^{-1}, \\ T_{n+1} &= \max[T_{\bar{\lambda}}, \min\{T_{\lambda}, T_n \\ &\quad + n^{-1}A_n(Y_n - p)\}], \end{aligned}\right\} \quad (n = 1, 2, \ldots). \tag{12}$$

THEOREM. *The SA plan is adaptive.*

Proof. It suffices to show that

$$T_n \to T_{\lambda} \text{ with probability one as } n \to \infty. \tag{13}$$

Writing $$W_n = n^{-1}A_n(Y_n - e^{-\lambda T_n}),$$

it follows from Theorem D, p. 387, Loève [58] that

$$\sum_1^n W_k \text{ converges with probability one as } n \to \infty. \tag{14}$$

Now suppose that for some $\omega \in \Omega$ for which (14) holds one can find a sequence $\{n_k\}$ such that

$$T_{n_k}(\omega) \to T_{\lambda} \quad \text{as} \quad k \to \infty \tag{15}$$

then it will be shown that

$$T_n(\omega) \to T_{\lambda} \quad \text{as} \quad n \to \infty. \tag{16}$$

Let $\epsilon > 0$ be small enough so that

$$(T_{\lambda} - \epsilon, T_{\lambda} + \epsilon) \subset (T_{\bar{\lambda}}, T_{\lambda}).$$

There exists an integer k_ϵ such that

$$|T_{n_k}(\omega) - T_{\lambda}| < \epsilon/2 \quad \text{for all} \quad k > k_\epsilon,$$

$$\left|\sum_m^{m+j} W_j\right| < \epsilon/4 \quad \text{for all} \quad m > n_{k_\epsilon} \quad \text{and all} \quad j > 0$$

and

$$(n_k + l)^{-1} A_{n_k+l} < \epsilon/4 \quad \text{for all} \quad k > k_\epsilon \quad \text{and all} \quad l > 0. \tag{17}$$

Let $k > k_\epsilon$ be fixed and consider the behaviour of $T_{n_k} + m$, $m = 0, 1, 2, \ldots$ suppose $T_{n_k} - T_{\lambda} > 0$ (a similar argument will hold otherwise). From (12),

$$\begin{aligned} T_{n_k+1} - T_{\lambda} &= T_{n_k} - T_{\lambda} + n_k^{-1}A_{n_k}(e^{-\lambda T_n k} - e^{-\lambda T_\lambda}) + W_{n_k} \\ &< T_{n_k} - T_{\lambda} + W_{n_k} < \epsilon. \end{aligned}$$

If $T_{n_k+1}-T_\lambda > 0$, then repeating,

$$T_{n_k+2}-T_\lambda \leqslant T_{n_k}-T_\lambda+W_{n_k}+W_{n_k+1} \leqslant \epsilon.$$

Repeating this argument one finds that if

$$T_{n_k+m}-T_\lambda > 0 \quad \text{for} \quad m = 0, 1, \ldots, l-1$$

then $$|T_{n_k+m}-T_\lambda| < \epsilon \quad \text{for} \quad m = 0, 1, \ldots, l-1.$$

Suppose that $T_{n_k+l}-T \leqslant 0$. Then

$$\begin{aligned} 0 &\geqslant T_{n_k+l}-T_\lambda \\ &= T_{n_k+l-1}-T_\lambda+(n_k+l-1)^{-1}A_{n_k+l-1}(e^{-\lambda T_{n_k+l-1}} \\ &\quad -e^{-\lambda T_\lambda})+W_{n_k}+l-1 \\ &> -(n_k+l-1)^{-1}A_{n_k+l-1}+W_{n_k+l-1} \\ &\geqslant -\epsilon/2. \end{aligned}$$

Now apply the same argument but starting with T_{n_k+l} instead of T_{n_k}. It follows that, for all m,

$$|T_{n_k+m}-T_\lambda| < \epsilon$$

and (16) follows.

In order to conclude the proof it will be shown now that for almost all ω, T_λ is a limit point of the sequence $\{T_n(\omega)\}$. Fix ω and let $\{T_{n_r}\}$ be a subsequence converging to T_0, say, with $T_0 \in [T_{\bar{\lambda}}, T_\lambda]$. Suppose that $T_0 > T_\lambda$, the other case being similar.

Let $\epsilon > 0$ be such that $T_\lambda+\epsilon < T_0-\epsilon$. There exists r_ϵ such that

$$|T_{n_r}-T_0| < \epsilon/2 \quad \text{for} \quad r > r_\epsilon,$$

$$\left|\sum_{m}^{m+j} W_i\right| < \epsilon/4 \quad \text{for} \quad m > n_{r\epsilon}, j > 0$$

and $$(n_r+l)^{-1}A_{n_r+l} < \epsilon/4 \quad \text{for} \quad 4 > r_\epsilon, l > 0. \tag{18}$$

Let $c = e^{-\lambda T_\lambda}-e^{-\lambda(T_\lambda+\epsilon)}$. Then, from (12) for $r > r_\epsilon$

$$T_{n_r+1}-T_\lambda \leqslant T_{n_r}-T_\lambda-Cn_r^{-1}A_{n_r}+W_{n_r}.$$

If $T_{n_r+1}-T_\lambda \geqslant \epsilon$, then one can repeat to obtain

$$T_{n_r+2}-T_\lambda \leqslant T_{n_r}-T_\lambda-C[n_2^{-1}A_{n_r}+(n_r+1)^{-1}A_{n_r+1}]+W_{n_r}+W_{n_r+1}.$$

If again $T_{n_k+2} - T_\lambda \geqslant \epsilon$, one can repeat as before: since

$$\sum_m (n_r + m)^{-1} A_{n_k+m} = \infty,$$

it follows that there exists an integer l such that

$$T_{n_r} + l - T_\lambda \leqslant \epsilon \quad \text{while} \quad T_{n_r+l-1} - T_\lambda \geqslant \epsilon.$$

Then, from (12)

$$T_{n_r+l} - T_\lambda \geqslant -(n_r + l - 1)^{-1} A_{n_r+l-1} + W_{n_r+l-1},$$

i.e.

$$|T_{n_r+l} - T_\lambda| \leqslant \epsilon.$$

Hence, for each ϵ there exists k such that $|T_k - T_\lambda| < \epsilon$, i.e. $\{T_n\}$ has a limit point at T_λ and the theorem is proved.

This is one of the illustrations of the way in which one can effectively apply stochastic-approximation techniques to solve applied problems. We take up this question again in the chapter on asymptotic normality and discuss its implications and utility.

5. Quantal response estimation

In the biological assay problem with quantal responses, an experimenter gives a selected dose of a drug to a laboratory animal. He can observe only one of two possible results for each animal, i.e., death or survival. Gaddum [40] presented a model for this experiment. For each animal, there is assumed to be a 'just-fatal dose' such that a larger dose would kill the animal, but a smaller dose would allow it to survive. Since animals vary, a frequency distribution $f(x)$ of the 'just-fatal dose' is assumed to exist. If a dose x is given the probability of killing an animal drawn at random is

$$F(x) = \int_{-\infty}^{x} f(y)\,dy. \tag{1}$$

The function $F(x)$, called the dose response curve, is a cumulative distribution function (*cdf*). The median lethal dose m is defined by $F(m) = \frac{1}{2}$. Traditionally the problem of estimation is attempted in the following non-sequential way.

If n animals are drawn independently and given the dose x, the probability that r die and s survive is the binomial term

$$\binom{n}{r}[F(x)]^r[1-F(x)]^s, \tag{2}$$

where $r+s=n$.

If J dose levels $X_1, \ldots, X_j$ are given to nJ animals with n animals at each dose level, the probability that $r_1, \ldots, r_j$ die and $s_1, \ldots, s_j$ survive is

$$\prod_{j=1}^{J}\binom{n}{r_j}[F(x_j)]^{r_j}[1-F(x_j)]^{s_j}, \tag{3}$$

where $r_j+s_j=n, j=1,2,\ldots,J$.

Then by the maximum likelihood method an estimate of m when the form of f is known is obtained or any other precentile can be estimated. But in this section we deviate from this practice and exploit the method of stochastic approximation and follow Venter [72]. Let us assume the following conditions on the density function.

(i) $f(y)$ is symmetric around $y=0$.

(ii) $f(y)$ is monotone non-increasing for $y \geqslant 0$.

(iii) $f(y)$ is twice differentiable for all real y.

Though these conditions are much stronger than required for the problem under consideration we shall need them in subsequent chapters.

Let m and τ be real numbers such that $0<\tau<\infty$. For each x (real), define

$$M(x) = M(x;\, m, \tau) = \frac{1}{\tau}\int_{-\infty}^{x} f\left(\frac{y-m}{\tau}\right)dy, \tag{4}$$

$$= \frac{1}{2} + \int_{0}^{(x-m)/\tau} f(y)\,dy, \tag{5}$$

where we have used (i) and the fact $\int_{-\infty}^{\infty} f(y)\,dy = 1$.

Further, for each x, let $Y(x)$ be a binomial random variable such that

$$P[Y(x)=1] = M(x) = 1-P[Y(x)=0]. \tag{6}$$

The problem of interest in this section is that of estimating m and τ, or functions of these parameters, given that it is possible to make observations on the random variable $Y(x)$ for any value of x. One can also find some material in Cochran and Davis [13 and 14] and in the references stated in their paper. We also discussed this situation at the beginning of Chapter 2 under the heading response-no-response analysis.

(*a*) *Median lethal dose estimation.*

Suppose τ is known and we want to estimate m. Then the following recursive relation will provide the result.

$$X_{n+1} = X_n - \frac{f(0)}{\tau n}[Y(x_n) - \tfrac{1}{2}] \tag{7}$$

since

$$M(x) - \frac{1}{2} = \frac{x-m}{\tau} f(0) + \frac{1}{\sigma}\left(\frac{x-m}{\tau}\right)^3 f''(0) + 0(|x-m|^3) \tag{8}$$

as $|x-m| \to 0$ from assumptions (i), (ii) and (iii). We can easily show that

$$(f'(0)) = 0, \tag{9}$$

$$E[Y(x) - M(x)]^2 = M(x)[1-M(x)] = \frac{1}{4} - \left[\int_0^{(x-m)/\tau} f(y)\,dy\right]^2 \tag{10}$$

and all the conditions of the Dvoretzky theorem hold so that $X_n \to m$ in mean square and with probability one as $n \to \infty$.

(*b*) *η-quantile effective dose estimation.*

Similarly we may take $0 < \eta < 1$ such that

$$\int_{-\infty}^{t} f(y)\,dy = \eta \tag{11}$$

and we can construct the Robbins–Monro procedure for estimating t. Thus a quantal response problem can in general be solved in the following way. Define m_η as the (unique) solution of the equation

$$M(x) = \eta. \tag{12}$$

Let

$$q_\eta = (m_\eta - m)/\tau; \tag{13}$$

using (5), q_η is uniquely determined by

$$\int_0^{\eta} f(y)\,dy = \eta - \tfrac{1}{2} \tag{14}$$

and therefore depends on $f(y)$ only. By (i) we have

$$m_\eta = 2m - m_{1-\eta} \quad \text{and} \quad q_\eta = -q_{1-\eta}. \tag{15}$$

It is often of interest to estimate m_η. With τ known, this can be done by estimating m as above and then using (13) to calculate an estimate of m_η.

However, since m_η is the solution of (12) and it is easy to check that all conditions of Robbins–Monro method hold, m_η can also be estimated directly by the following recursive relation.

$$X_{n+1} = X_n - \frac{1}{\tau n} f(q_\eta)\,[Y(x_n) - \eta].$$

We shall take up this problem again for consideration of asymptotic distribution. It is obvious from all that went before that the Robbins–Monro method is very useful in problems of bioassay. Davis [18] has considered application of the Robbins–Monro method, and the Kesten version of the Robbins–Monro method, with their modification to problems of Bioassay, and made their comparison. Wetherill points out that for estimation of the 90th percentile or so this method is not efficient, see Wetherill [82] and Guttman and Guttman [43].

CHAPTER 5

MULTIVARIATE STOCHASTIC-APPROXIMATION METHODS

1. Introduction

Multidimensional techniques of stochastic approximation have become very popular due to their practical utility and simplicity. For example, the problems of the improvement of system performance by experimental techniques are becoming increasingly important. In the construction of systems there are many problems which often force one to be satisfied with a design that is less than optimum. The theory may not be well understood; the effects of certain factors may not be known and so forth. Under these conditions, it is often desired, once a reasonable system has been built, to attempt to optimize its performance by experimenting with the values of its adjustable parameters. One would take measurements of the performance at various parameter settings and use the acquired information to obtain better parameter settings. There are two factors which complicate the interpretation of experimental data. The first is the presence of 'noise', the second is the ignorance of the form of the functional dependence of the performance upon the parameters. The iterative processes for experimental optimization have a great deal of interest, since their use necessitates neither the direct estimation of the unknown parameters nor the observation of the inputs. Thus many engineers have been attracted to use these techniques in designing and improving various systems. We shall take up this matter again in Chapter 7, where we discuss stochastic-approximation techniques for continuous random processes.

2. Multivariate Robbins–Monro method

In this section we discuss results of Blum [5] who has designed multidimensional stochastic-approximation techniques and stated conditions under which they converge almost surely to the solution of k stochastic equations in k unknowns and to the point where a regression function in k variables achieves its maximum.

THEOREM. 1 (Blum). Let $\{Y^{(1)}_{X_1,\ldots,X_k}\},\ldots,\{Y^{(k)}_{X_1,\ldots,X_k}\}$ be k families of random variables with corresponding families of distribution functions $\{F^{(1)}_{X_1,\ldots,X_k}\},\ldots,\{F^{(k)}_{X_1,\ldots,X_k}\}$, each depending on k real variables $(X_1,\ldots,X_k)$. Let $M^i(X_1,\ldots,X_k)=\int_{-\infty}^{\infty} y\,dF^i_{X_1,\ldots,X_k}$, for $i=1,\ldots,k$ be the corresponding regression functions. It is assumed that the distributions $F^i_{X_1,\ldots,X_k}$ and $M^i(X_1,\ldots,X_k)$, $i=1,\ldots,k$ are unknown; however, it is possible to make an observation on the random variable $Y^i_{X_1,\ldots,X_k}$ for $i=1,2,\ldots,k$ and any choice of real numbers $(X_1,\ldots,X_k)$.

Let R^k be a real k-dimensional vector space spanned by the orthogonal unit vectors $\mathbf{u}_1,\ldots,\mathbf{u}_k$. If $\mathbf{x}$ and $\mathbf{y}$ are two vectors in R^k, we denote their inner product by $\langle\mathbf{x},\mathbf{y}\rangle$ and their norms by $\|\mathbf{x}\|$ and $\|\mathbf{y}\|$ respectively. Suppose that to each $\mathbf{x}\in R^k$ corresponds a random vector $\mathbf{Y}_x\in R^k$. Denote by $\mathbf{M}(\mathbf{x})$ the vector representing the conditional expectation of $\mathbf{Y}_x$ when $\mathbf{X}$ is fixed.

Let now $f(\mathbf{x})$ be a real-valued function defined on R^k and possessing continuous partial derivatives of the first and second order. The vector of first partial derivatives will be denoted by $\mathbf{D}(\mathbf{x})$ and the matrix of second partial derivatives by $\mathbf{A}(\mathbf{x})$. That is,

$$\mathbf{D}(\mathbf{x})=\left(\frac{\partial f}{\partial x_i}\right)\Bigg|_{\mathbf{x}},\quad \mathbf{A}(\mathbf{x})=\left(\frac{\partial^2 f}{\partial x_i\,\partial x_j}\right)\Bigg|_{\mathbf{x}}.$$

Then, for any real number a, we have, by Taylor's theorem,

$$f(\mathbf{x}+a\mathbf{Y}_x)=f(\mathbf{x})+a\langle\mathbf{D}(\mathbf{x}),\mathbf{Y}_x\rangle+\tfrac{1}{2}a^2\langle\mathbf{Y}_x,\mathbf{A}(x+\theta a\mathbf{Y}_x)\,\mathbf{Y}_x\rangle,$$

where θ is a real number with $0\leqslant\theta\leqslant 1$. If expectation is taken on both sides one obtains

$$E\{f(\mathbf{x}+a\mathbf{Y}_x)\}=f(\mathbf{x})+a\langle\mathbf{D}(\mathbf{x}),\mathbf{M}(\mathbf{x})\rangle + \tfrac{1}{2}a^2E\{\langle\mathbf{Y}_x, A(\mathbf{x}+\theta a\mathbf{Y}_x)\mathbf{Y}_x\rangle\}.\qquad(1)$$

Let now $\{a_n\}$ be a sequence of positive numbers and consider the following sequence of recursively defined random vectors.

$$\mathbf{X}_{n+1}=\mathbf{X}_n+a_n\mathbf{Y}_n,\qquad(2)$$

where $\mathbf{X}_1$ is chosen arbitrarily and where $\mathbf{Y}_n$ has the distribution of $\mathbf{Y}_x$ when $\mathbf{X}_n$ yields the observation $\mathbf{X}$. Let us employ the following notation.

$$Z_x = f(\mathbf{x}),\ U(\mathbf{x}) = \langle \mathbf{D}(\mathbf{x}), \mathbf{M}(\mathbf{x})\rangle,$$
$$V_a(\mathbf{x}) = E\{\langle \mathbf{Y}_x, \mathbf{A}(\mathbf{x}+\theta a\mathbf{Y}_x)\,\mathbf{Y}_x\rangle\}$$

when the random variables $\mathbf{X}_n$ for $\mathbf{x}$ and numbers a_n for a are substituted the corresponding random variables are denoted by Z_n, U_n and V_n.

Assume $\mathbf{M}(\mathbf{0}) = \mathbf{0}$ without loss of generality, and consider the following set A of conditions.

(i) $\sum_1^\infty a_n = \infty$, $\sum_1^\infty a_n^2 < \infty$,

(ii) $Z_x \geqslant 0$.

(iii) $\sup\limits_{\epsilon < \|\mathbf{x}\|} U(\mathbf{x}) < 0$ *for every* $\epsilon > 0$.

(iv) $\inf\limits_{\epsilon \leqslant \|\mathbf{x}\|} \|Z_x - Z_0\| > 0$ *for every* $\epsilon > 0$.

(v) $V_a(\mathbf{x}) < V < \infty$ *and* $V \geqslant 0$ *for every number* a.

Then the sequence $\{\mathbf{X}_n\}$ *defined by* (2) *converges to zero almost surely* (*a.s.*).

Proof. From (1) one obtains,

$$E(Z_{n+1}|Z_1, \ldots, Z_n) = Z_n + a_n E\{U_n|Z_1, \ldots, Z_n\} + \frac{a_n^2}{2} E\{V_n|Z_1, \ldots, Z_n\} \text{ a.s.} \quad (3)$$

Since $\mathbf{M}(\mathbf{0}) = \mathbf{0}$, we have, by virtue of conditions A,

$$E[U_n|Z_1, \ldots, Z_n] \leqslant a \text{ a.s.}, \quad E[V_n|Z_1, \ldots, Z_n] \leqslant V \text{ a.s.},$$

both for all n. Otherwise one can easily prove a contradiction to conditions A. Hence

$$E[Z_{n+1} - Z_n|Z_1, \ldots, Z_n] \leqslant \tfrac{1}{2}a_n^2 V \text{ a.s.} \quad (4)$$

By conditions (i) and (ii) and problem 2 [Appendix II, §4] one obtains

$$P[Z_n \text{ converges}] = 1. \quad (5)$$

Taking expectations on both sides of (3) and iterating one obtains

$$E(Z_{n+1}) = Z_1 + \sum_{j=1}^n a_j E\{U_j\} + \sum_{j=1}^n \tfrac{1}{2}a_j^2 E[V_j].$$

From what has been said above and from the property of expectation of conditional expectation it follows that

$$E\{Z_n\} \geqslant 0, \quad E\{U_n\} \leqslant 0, \quad E\{V_n\} \leqslant V \quad (n = 1, 2, \ldots).$$

Since V is non-negative and the series $\sum_1^\infty a_n^2$ converges, the non-positive term series $\sum_1^\infty a_n E(U_n)$ also converges. By virtue of the fact that $\sum_1^\infty a_n = \infty$ one has

$$\limsup_{n\to\infty} E\{U_n\} = 0, \quad \liminf_{n\to\infty} E\{|U_n|\} = 0.$$

Let $\{n_r\}$ be an infinite sequence of integers such that

$$\lim_{r\to\infty} E\{|U_{n_r}|\} = 0.$$

Then U_{n_k} converges to zero in probability due to Tchebichev's inequality and there exists a further subsequence say $\{U_{m_r}\}$ such that

$$P\{\lim_{r\to\infty} U_{m_r} = 0\} = 1.$$

From condition (iii) it follows that $P[\lim_{r\to\infty} X_{m_r} = 0] = 1$. Since Z_n is a continuous function of $\mathbf{X}_n$ it follows from (5) that

$$P[\lim_{n\to\infty} Z_n = Z_0] = 1. \tag{6}$$

Now consider a sample sequence $\{\mathbf{X}_n\}$ such that for the corresponding sequence $\{Z_n\}$ one has limit $Z_n = Z_0$. From condition (iv) it is obvious that for such a sequence one must have

$$\lim_{n\to\infty} \mathbf{X}_n = \mathbf{0};$$

otherwise it will lead to a contradiction of (iv). Hence (6) gives the desired result.

Thus if $(\alpha_1, \ldots, \alpha_k) = \boldsymbol{\alpha}$ are k specified numbers then

$$\mathbf{X}_{n+1} = \mathbf{X}_n + a_n(\mathbf{Y}_n - \alpha) \tag{1''}$$

is, assuming all conditions stated, a stochastic-approximation procedure which gives the solution of the set of equations

$$M^i(X_1, \ldots, X_k) = \alpha_i \quad (i = 1, \ldots, k)$$

with probability one.

Now an illustrative example is discussed in order to show that the conditions of the theorem can be met.

Example (Blum). Let $\mathbf{B}$ be a negative definite $k \times k$ matrix and assume

(i′) for some $\rho > 0 \|\mathbf{x}\| \leqslant \rho$ implies $\mathbf{M}(\mathbf{x}) = \mathbf{B}\mathbf{x}$.

(ii′) $\|\mathbf{x}\| > \rho$ implies $\mathbf{M}(\mathbf{x}) = \mathbf{M}([\rho/\|\mathbf{x}\|]\,\mathbf{x})$.

(iii′) $\sigma_{\mathbf{x}}^{2(i)} < \sigma^2 < \infty$ for each $\mathbf{x} \in R^k$ and each $i = 1, \ldots, k$, where $\sigma_{\mathbf{x}}^{2(i)}$ is the variance of the ith component of $\mathbf{Y}_x$.

Thus $\|\mathbf{M}(\mathbf{x})\|$ and $E\{\|\mathbf{Y}_{\mathbf{x}}^2\|\}$ are bounded uniformly in $\mathbf{x}$. Let now $f(\mathbf{x}) = \|\mathbf{x}\|^2$ and choose a sequence $\{a_n\}$ satisfying (i). (ii) and (iv) are obviously satisfied for choice of $f(\mathbf{x}) = \|\mathbf{x}^2\|$ and further we have

$$U(\mathbf{x}) = \begin{cases} 2\langle \mathbf{x}, \mathbf{B}\mathbf{x}\rangle & \|\mathbf{x}\| \leqslant \rho, \\ 2[\rho/\|\mathbf{x}\|]\langle \mathbf{x}, \mathbf{B}\mathbf{x}\rangle & \|\mathbf{x}\| > \rho, \end{cases}$$

$$V_a(\mathbf{x}) = 2E\{\|\mathbf{Y}_x^2\|\} \quad \text{for every number } a.$$

From boundedness of $E\{\|\mathbf{Y}_x^2\|\}$ it is clear that (v) is also satisfied. In order to check (iii) one can use the fact that for every negative definite matrix B there exists a positive number b such that $\langle \mathbf{x}, \mathbf{B}\mathbf{x}\rangle \leqslant -b\|\mathbf{x}^2\|$. Thus if ϵ is any positive number with $0 < \epsilon < \rho$ then $\langle \mathbf{x}, \mathbf{B}\mathbf{x}\rangle \leqslant -b\epsilon^2$ if $\epsilon \leqslant \|\mathbf{x}\| \leqslant \rho$,

$$[\rho/\|\mathbf{x}\|]\langle \mathbf{x}, \mathbf{B}\mathbf{x}\rangle \leqslant -b\rho^2 \quad \text{if} \quad \|\mathbf{x}\| > \rho.$$

Hence (iii) is also satisfied. Hence all conditions of the theorem are met for this simple example, so that one can make use of Theorem 1 in this case.

In the next section we discuss a situation where one can reduce the multidimensional stochastic-approximation case to one-dimensional stochastic approximation and milk out a result from the Dvoretzky theorem. This is discussed in Epling [30]

3. Application to a problem of pharmacology

In § 3 of Chapter 4 we discussed a kinetic model in pharmacology where two drugs are competing. Now we look at a kinetic model where k drugs are involved. The problem can be formulated mathematically in the following way.

(*a*) *Mathematical formulation.*

Suppose we have k families of random variables

$$\{Y_i(x)\} \quad i = 1, 2, \ldots, k$$

with corresponding distribution function

$$H_i(Y|X), \quad M_i(x) = \int_{-\infty}^{\infty} Y dH_i(Y|X)$$
$$\text{and} \quad \sigma_i^2(x) = E[Y - M_i(x)]^2.$$

Further suppose that the k regression functions $M_i(x)$ satisfy the following conditions.

(i) There exists $\mathbf{\theta}_s = [\theta_1, \ldots, \theta_k]$ uniquely determined by $\sum_1^k \theta_i = s$.

(ii) $M_i(\theta_i) = M_j(\theta_j) \quad (i, j = 1, 2, 3, \ldots, k)$, (1)

where the M's are strictly monotone increasing, continuous and the intersection of their ranges is an interval of positive length.

Let

$$a_n > 0, \quad \sum_1^\infty a_n = \infty, \quad \text{and} \quad \sum_1^\infty a_n^2 < \infty. \tag{2}$$

Sequential sets of k experiments are to be performed in the following manner. A vector of initial levels

$$\mathbf{X}_1 = [X_{11}, X_{21}, \ldots, X_{k1}]$$

is chosen arbitrarily and succeeding levels for each set of experiments are defined by

$$\mathbf{X}_{n+1} = \mathbf{X}_n + a_n[\overline{\mathbf{Y}}_n - \mathbf{Y}_n], \tag{3}$$

where
$$\mathbf{X}_n = [X_{1n}, X_{2n}, \ldots, X_{kn}],$$
$$\mathbf{Y}_n = [Y_{1n}, Y_{2n}, \ldots, Y_{kn}],$$
$$\overline{\mathbf{Y}}_n = [\overline{Y}_n, \overline{Y}_n, \ldots, \overline{Y}_n],$$

where
$$\overline{Y}_n = \frac{1}{k}\sum_1^k Y_{in}.$$

It is assumed that the k experiments in each set are performed independently; in other words the only dependence among the experiments is that arising from the recursive selection of the levels at which they are conducted.

Suppose for all x and $i = 1, \ldots, k$ there exist V, A and B, positive real numbers, such that

$$\sigma_i^2(x) < V < \infty, \tag{4}$$

$$|M_i(x)| < A|x| + B < \infty. \tag{5}$$

THEOREM 2. *Under the above-mentioned conditions*

$$E\{\|\mathbf{X}_n - \boldsymbol{\theta}_s\|^2\} \to 0 \quad as \quad n \to \infty \tag{6}$$

and $\|\mathbf{X}_n - \boldsymbol{\theta}_s\| \to 0$ *as* $n \to \infty$ *with probability one.*

Proof. Choose
$$\mathbf{X}_1 = [X_{11}, X_{21}, \ldots, X_{k1}]$$

such that
$$\sum_1^k X_{i1} = s$$

and let $X_1' = X_i - \theta_i$ and $Y_i(x_i) - M_i(\theta_i) = Y_i'(x')$.

Then convergence of the x_i' to zero using the families of random variables $\{Y_i'(x_i')\}$ is equivalent to convergence of the X_i to θ_i using the given families. Thus without loss of generality we may take $\boldsymbol{\theta}_s$ to be the null vector and $M_i(\theta_i)$ to be zero for $i = 1, \ldots, k$. Thus we have to prove that (we shall omit prime).

$$E\{\|X_n\|^2\} \to 0 \quad \text{as} \quad n \to \infty, \tag{7}$$

$\|X_n\| \to 0$ as $n \to \infty$ with probability one. Further

since
$$\sum_1^k \theta_i = 0, \quad \sum_1^k X_{in} = 0 \quad \text{for all } n.$$

Let
$$\mathbf{M}_n = [M_1(x_{1n}), \ldots, M_1(x_{1n})],$$

$$\overline{\mathbf{M}}_n = [\overline{M}_1, \ldots, \overline{M}_n],$$

where
$$\overline{M}_n = \frac{1}{k}\sum_1^k M_i(x_{in}).$$

Let
$$\mathbf{Z}_n = [\overline{\mathbf{Y}}_n - \mathbf{Y}_n] - [\overline{\mathbf{M}}_n - \mathbf{M}_n].$$

Now $\sum_1^k Z_{in} = 0$, where $Z_n = [Z_{1n}, Z_{2n}, Z_{3n}, \ldots, Z_{kn}]$

and then the sequence (3) can be written

$$\mathbf{X}_{n+1} = \mathbf{X}_n + a_n(\overline{\mathbf{M}}_n - \mathbf{M}_n) + a_n \mathbf{Z}_n,$$

we have

$$\|\mathbf{X}_{n+1} - a_n \overline{\mathbf{M}}_n\|^2 = \|\mathbf{X}_n - a_n \mathbf{M}_n + a_n \mathbf{Z}_n\|^2.$$

The inner product of $\mathbf{X}_{n+1}$ and $\overline{\mathbf{M}}_n$ vanishes since $\sum_1^k Z_{in+1} = 0$ and all components of $\overline{\mathbf{M}}_n$ are identical, by the same argument the inner product of $\mathbf{Z}_n$ and $\overline{\mathbf{M}}_n$ vanishes. Thus

$$\|\mathbf{X}_{n+1}\|^2 = \|\mathbf{X}_n - a_n \overline{\mathbf{M}}_n + a_n \mathbf{Z}_n\|^2 - a_n^2 k \overline{\mathbf{M}}_n^2.$$
$$\leqslant \|\mathbf{X}_n - a_n \mathbf{M}_n + a_n \mathbf{Z}_n\|^2.$$

Therefore

$$E\{\|\mathbf{X}_{n+1}\| \,|\mathbf{X}_n\} \leqslant E\{\|\mathbf{X}_n - a_n \mathbf{M}_n + a_n \mathbf{Z}_n\| \,|\mathbf{X}_n\}.$$

Let
$$r_n = \|\mathbf{X}_n\| \tag{8}$$

and define the random transformation

$$t_n(r_n) = E\{r_{n+1}|\mathbf{x}_n\}. \tag{9}$$

It remains only to show that the sequences $\{r_n\}$ and transformations $t_n(r)$ satisfy the conditions of the Dvoretzsky theorem. Let

$$Z_n = r_{n+1} - t_n(r_n). \tag{9a}$$

It will be shown that for any $n \geqslant 1$, $E\{Z_n^2\} < 4a_n^2 kV$; hence $\sum_{n=1}^{\infty} E(Z_n^2) < \infty$.

$$E(Z_n^2) = E\{[r_{n+1} - E\{r_{n+1}|\mathbf{x}_n\}]^2 \,|\mathbf{x}_n\}$$
$$= E\{[\|\mathbf{x}_{n+1}\| - E\{\|\mathbf{x}_{n+1}\| \,|\mathbf{x}_n\}]^2 \,|\mathbf{x}_n\}.$$

On the one hand,

$$\|\mathbf{x}_{n+1}\| = \|\mathbf{x}_n + a_n(\overline{\mathbf{M}}_n - \mathbf{M}_n) + a_n \mathbf{z}_n\|$$
$$\leqslant \|\mathbf{x}_n + a_n(\overline{\mathbf{M}}_n - \mathbf{M}_n)\| + \|a_n \mathbf{z}_n\|,$$

and on the other

$$\|\mathbf{x}_{n+1}\| \geqslant \|\mathbf{x}_n + a_n(\overline{\mathbf{M}}_n - \mathbf{M}_n)\| - \|a_n \mathbf{z}_n\|.$$

Therefore,

$$\|\mathbf{x}_n + a_n(\overline{\mathbf{M}}_n - \mathbf{M}_n)\| + E\{\|a_n \mathbf{z}_n\| \,|x_n\}$$
$$\geqslant E\{\|\mathbf{x}_{n+1}\| \,|\mathbf{x}_n\}$$
$$\geqslant \|\mathbf{x}_n + a_n(\overline{\mathbf{M}}_n - \mathbf{M}_n)\| - E\{\|a_n \mathbf{z}_n\| \,|\mathbf{x}_n\},$$

and consequently

$$\begin{aligned}
&E\{[\|\mathbf{x}_{n+1}\| - E\{\|\mathbf{x}_{n+1}\| \,|\mathbf{x}_n\}]^2\} \\
&\quad \leqslant E\{[\|a_n \mathbf{z}_n\| + E\{\|a_n \mathbf{z}_n\| \,|\mathbf{x}_n\}]^2 \,|\mathbf{x}_n\} \\
&\quad \leqslant 2E\{\|a_n \mathbf{z}_n\|^2 |\mathbf{x}_n\} + 2E^2\{\|a_n \mathbf{z}_n\| \,|\mathbf{x}_n\} \\
&\quad \leqslant 4a_n^2 kV. \qquad (9a.1)
\end{aligned}$$

It remains only to establish sequences α_n and $\gamma_n(r)$, fulfilling the desired conditions.

$$\begin{aligned}
|t_n(r_n)| &= E\{\|\mathbf{x}_{n+1}\| \,|\mathbf{x}_n\} \\
&\leqslant E\{\|\mathbf{x}_n - a_n \mathbf{M}_n + a_n \mathbf{z}_n\| \,|\mathbf{x}_n\} \\
&\leqslant \Big[\|\mathbf{x}_n - a_n \mathbf{M}_n\|^2 + a_n^2 kV \\
&\qquad + 2 \sum_{i=1}^{k} (x_{in} - a_n M_{in}) E\{z_{in} |\mathbf{x}_n\} \Big]^{\frac{1}{2}} \\
&= [\|\mathbf{x}_n - a_n \mathbf{M}_n\|^2 + a_n^2 kV]^{\frac{1}{2}}. \qquad (9b)
\end{aligned}$$

Next, various bounds depending on n and $\|\mathbf{x}_n\|$ will be established; then the term $a_n^2 kV$, which goes to zero relatively rapidly, will be combined with these bounds to give bounds, depending on n and r_n on $|t_n(r_n)|$. Consider $\|\mathbf{x}_n - a_n \mathbf{M}_n\|^2$.

Choose a positive sequence ρ_n, $n \geqslant 1$, such that

$$\sum_{n=1}^{\infty} a_n \rho_n = \infty.$$

Choose another positive sequence, η_n, such that

$$\inf_{\eta_n \leqslant |x| \leqslant 1} |M_i(x)| > \rho_n \quad \text{for} \quad i = 1, \ldots, k,$$

and
$$a_n/\rho_n \eta_n \to 0 \quad \text{and} \quad a_n \rho_n / 2k\eta_n < 1.$$

This is easily done since the strictly monotone increasing properties of the $M_i(x)$ guarantee the existence of some sequence, η_n, fulfilling the first condition; further, any sequence uniformly greater than η_n will also fulfill the first condition. A sequence $\{a_n\}$ having property A is asymptotically less than $n^{-\frac{1}{2}}$; hence choose $\rho_n = n^{-\frac{1}{8}}$, and $\eta_n > n^{-\frac{1}{8}}$, say, and the latter conditions are fulfilled also. Since x and $M_i(x)$ have the same sign,

$$|x - a_n M_i(x)| = \left| |x| - a_n |M_i(x)| \right|,$$

and
$$|x| - a_n |M_i(x)| \leqslant |x| - a_n \rho_n$$

for
$$|x| > \eta_n \quad \text{and} \quad i = 1, \ldots, k.$$

Thus if $x - a_n M_i(x)$ has the same sign as x, then $|x| - a_n|M_i(x)| > 0$ and we can write

$$|x - a_n M_i(x)| \leqslant (1 - a_n \rho_n / |x|)\, |x| \tag{9c}$$

for
$$\eta_n < |x| \quad \text{and} \quad i = 1, \ldots, k.$$

On the other hand, if $x - a_n M_i(x)$ does not have the same sign as x, $|x| - a_n |M_i(x)| < 0$ and condition (5) guarantees that there exists n_A, say, such that for $n > n_A$

$$|x| - a_n |M_i(x)| \geqslant -a_n B \quad \text{for all} \quad x, i = 1, \ldots, k. \tag{9d}$$

Combining (9c) and (9d), one obtains, for $n > n_A$

$$|x - a_n M_i(x)|^2 \leqslant (1 - a_n \rho_n / |x|)^2\, |x|^2 + a_n^2 B^2$$

for
$$|x| > \eta_n \quad \text{and for} \quad i = 1, \ldots, k,$$

$$(1 - a_n \rho_n / |x|)^2 = [1 - a_n \rho_n (2/|x| - a_n \rho_n / |x|^2)].$$

For
$$n > n_0, \quad \text{say, and} \quad |x| > \eta_n$$

$$2/|x| - a_n \rho_n / |x|^2 > 1/|x| + (1/|x|)\,(1 - a_n \rho_n / \eta_n) > 1/|x|.$$

Since ρ_n and η_n were so chosen that $a_n \rho_n / \eta_n \to 0$. Hence,

$$|x - a_n M_i(x)|^2 \leqslant (1 - a_n \rho_n / |x|)\, |x|^2 + a_n^2 B^2 \tag{9e}$$

for
$$n > \max(n_A, n_0),$$

$$|x| > \eta_n \quad \text{and} \quad i = 1, \ldots, k.$$

By an argument similar to that used in developing (9e), one can establish an asymptotic bound for $|x - M_i(x)|$ for all x. Since x and $M_i(x)$ have the same sign, $|x| - a_n |M_i(x)| < |x|$ for all x, n, $i = 1, \ldots, k$. This in conjunction with (9d) gives

$$|x - M_i(x)|^2 \leqslant |x|^2 + a_n^2 B^2 \quad \text{for all} \quad x, n \geqslant n_A. \tag{9f}$$

Now (9e) and (9f) will be applied to $\|\mathbf{x}_n - a_n \mathbf{M}_n\|^2$. For any vector $\mathbf{x} = [x_1, \ldots, x_k]$, letting $x_{\max}$ be a component satisfying

$$|x_{\max}| = \max[|x_1|, \ldots, |x_k|]$$

and letting $0 < C < 1$,

$$\|x\|^2(1 - C/k) = (1 - C/k) \sum_{i=1}^{k} |x_i|^2$$
$$\geqslant \sum_{i=1}^{k} |x_i|^2 + (1 - C)|x_{\max}|^2,$$
$$|x_i| \neq |x_{\max}|.$$

If $$k\eta_n < |r_n| = \|\mathbf{x}_n\|,$$

then $$\eta_n < |x_{\max}|$$

and therefore for $$n > \max(n_A, n_0),$$

$$\|\mathbf{x}_n - a_n \mathbf{M}_n\|^2 \leqslant \sum_{\substack{i=1 \\ x_i \neq x_{\max}}}^{k} |x_{in} - a_n M_i(x_{in})|^2$$
$$+ (1 - a_n \rho_n / |x_{\max}|)\, |x_{\max}|^2 + k a_n^2 B^2$$
$$\leqslant \sum_{\substack{i=1 \\ x_i \neq x_{\max}}}^{k} |x_{in} - a_n M_i(x)|^2 + (1 - a_n \rho_n / \|\mathbf{x}_n\|)|x_{\max}|^2$$
$$\leqslant (1 - a_n \rho_n / \|x_n\|)\|\mathbf{x}_n\|^2.$$

Thus, from (9b),

$$|t_n(r_n)|^2 \leqslant (1 - a_n \rho_n / k|r_n|)|r_n|^2 + k a_n^2 (B^2 + V)$$
$$= |r_n|^2 [1 - a_n \rho_n (1/k|r_n| - k a_n (B^2 + V)/\rho_n |r_n|^2)]$$

for the range of $|r_n|$ and n specified above. As before, the coefficient in the above equation can be somewhat simplified.

$$1/k|r_n| - k a_n (B^2 + V)/\rho_n |r_n|^2$$
$$> 1/2k|r_n| + (1/2k|r_n|)(1 - 2k a_n (B^2 + V)/\rho_n \eta_n)$$
$$> 1/2k|r_n|$$

for $n > n_1$, say, and $|r_n| > k\eta_n$; thus

$$|t_n(r_n)|^2 \leqslant (1 - a_n \rho_n / 2k|r_n|)|r_n|^2 \tag{9g}$$

for sufficiently large n and $|r_n| > k\eta_n$.

Applying (9*f*) to (9*b*), one obtains for $n > n_A$

$$|t_n(r_n)|^2 \leqslant |r_n|^2 + ka_n^2(B^2+V) \tag{9h}$$

for all x. In particular,

$$|t_n(r_n)|^2 \leqslant k^2\eta_n^2 + ka_n^2(B^2+V) \tag{9i}$$

for sufficiently large n and $|r_n| < k\eta_n$.

Note that (9*g*), (9*h*) and (9*i*) hold uniformly for the random transformation t_n. Suppose the desired sequences, α_n and $\gamma_n(r)$ are defined as follows:

$$\alpha_n = [k^2\eta_n^2 + ka_n^2(B^2+V)]^{\frac{1}{2}}$$

and

$$\gamma_n(r) = \begin{cases} 1-[1-a_n\rho_n/2k|r|]^{\frac{1}{2}} & \text{if } \; a_n\rho_n/2k|r| < 1, \\ 1 \text{ otherwise.} \end{cases}$$

Each α_n is non-negative and $\lim_{n\to\infty} \alpha_n = 0$. Each $\gamma_n(r)$ is a non-negative function and for any sequence $r_1, \ldots, r_n, \ldots$ such that

$$\sup_{n=1,2,\ldots} |r_n| < L < \infty, \quad \gamma_n(r_n) > \gamma_n(L) > a_n\rho_n/2kL.$$

Let $N = \max(n_A, n_0, n_1)$. For $n > N$, either $|r| \leqslant k\eta_n$ in which case $|t_n(r)| \leqslant \alpha_n$ by (9*i*), or $|r| > k\eta_n$ in which case $|t_n(r)| < (1-\gamma_n(r))|r|$ by (9*g*) and construction of the sequence η_n. Thus for $n > N$, condition 4 of the Dvoretzky Theorem 3 holds using α_n and $\gamma_n(r)$ as defined above.

If condition (4) held for all n, then by the Dvoretzky Theorem (3) the result would have been proved. The only point remaining to complete the proof of the result is to show that for the stochastic-approximation process considered here it is sufficient for (4) to hold only for large n. Suppose all sequences (of random variables, constants, functions, transformations, etc.) are renumbered from the $(N+1)$st step on: i.e. let $r_{(m)} = r_{n-N}$, $t_{(m)}(r) = t_{n-N}(r)$, etc. for $n > N$. Then it is obvious that the asymptotic behaviour of the renumbered sequences is identical to the asymptotic behaviour of the original sequences; moreover it is obvious that all of the hypotheses of Dvoretzky's theorem are fulfilled for the 'new' collection of sequences, except possibly condition (7) of Dvoretzky's Theorem (3) that

$$E\{r_{(1)}^2\} = E\{r_{N+1}^2\} < \infty.$$

From (9), (9a) and (9a.1) we have, for all n, that

$$E\{r_{n+1}^2\} \leqslant E\{(t_n(r_n))^2\} + 4a_n^2 kV$$

and from (9h) we have

$$E\{r_{n+1}^2\} \leqslant E\{r_n^2\} + ka_n^2(B^2+V) + 4a_n^2 kV.$$

Thus by recursion, we have that

$$E\{r_{N+1}^2\} \leqslant E\{r_1^2\} + k(B^2+5V) \sum_{n=1}^{N} a_n^2 < \infty.$$

Hence all the conditions of the Dvoretzky theorem are satisfied, and hence the result follows.

4. Multivariate Kiefer–Wolfowitz method

In this section we consider a stochastic-approximation method (which is the content of Theorem 3) by which one can locate the maximum of a regression function in several variables.

THEOREM 3. (Blum). *We follow the notation of Theorem* 1.

Let $\mathbf{x}$ *be a variable point in* R^k *and suppose that to each* $\mathbf{x}$ *corresponds a random variable* $Y_\mathbf{x}$ *with corresponding regression function* $M(\mathbf{x})$. *Assume without loss of generality that* $M(\mathbf{x})$ *has a unique maximum at* $\mathbf{x} = 0$.

Let $\{a_n\}$ and $\{c_n\}$ be two sequences of positive numbers satisfying the following set of conditions.

(i) $\lim_{n\to\infty} c_n = 0$, (ii) $\sum_1^\infty a_n = \infty$,

(iii) $\sum_1^\infty a_n c_n < \infty$, (iv) $\sum_1^\infty (a_n/c_n)^2 < \infty$.

Let c be a positive number and let $\mathbf{u}_1, \ldots, \mathbf{u}_k$ be an orthonormal set spanning R^k. We construct a random vector $Y_{\mathbf{x},c}$ by taking $(k+1)$ independent observations on the random variables $Y_\mathbf{x}, Y_{\mathbf{x}+c\mathbf{u}_1}, \ldots, Y_{\mathbf{x}+c\mathbf{u}_k}$ and defining

$$\mathbf{Y}_{\mathbf{x},c} = [(Y_{\mathbf{x}+c\mathbf{u}_1} - Y_\mathbf{x}), \ldots, (Y_{\mathbf{x}+c\mathbf{u}_k} - Y_\mathbf{x})].$$

We proceed to construct a recursive sequence of random vectors by choosing $\mathbf{X}_1$ arbitrarily and defining

$$\mathbf{X}_{n+1} = \mathbf{X}_n + a_n Y_n / c_n, \tag{1}$$

where $\mathbf{Y}$ has the distribution of $\mathbf{Y}_{\mathbf{x}, c_n}$ when $\mathbf{X}_n$ yields the observation $\mathbf{X}$. $\mathbf{Y}_n/c_n$ is the vector in the direction of the maximum slope of the plane determined by the $k+1$ vectors

$$(\mathbf{X}_n, \mathbf{Y}_{x_n}), (\mathbf{X}_n + c_n \mathbf{u}_1, \mathbf{Y}_{\mathbf{x}_n + c_n \mathbf{u}_1}), \ldots, (\mathbf{X}_n + c_n \mathbf{u}_k, \mathbf{Y}_{\mathbf{x}_n + c_n \mathbf{u}_k}).$$

Assume that

(i) $M(\mathbf{x})$ is continuous with continuous first and second derivatives; we denote the vector of first partial derivatives and matrix of second partial derivatives of $M(\mathbf{x})$ by $\mathbf{D}(\mathbf{x})$ and $\mathbf{A}(\mathbf{x})$ respectively;

(ii) the second partial derivatives $\partial^2 M / \partial x_i \partial x_j$ are bounded for $i, j = 1, \ldots, k$.

We denote by $\bar{\mathbf{A}}_n$ the vector whose coordinates are diagonal entries of $\mathbf{A}_n$, and by $\mathbf{\Delta}_n$ the vector $E\{\mathbf{Y}_n | \mathbf{x}_n\}$.

(iii) We denote by σ_x^2 the variance of Y_x and $\sigma_x^2 \leqslant \sigma^2 < \infty$ and without loss of generality $M(\mathbf{0}) = \mathbf{0}$ so that $M(\mathbf{x}) < \mathbf{0}$ for all $\mathbf{x}$. Assume also that

(iv) For every positive number ϵ there exists a positive number $\rho(\epsilon)$ such that

$$\|\mathbf{x}\| \geqslant \epsilon \quad \text{implies} \quad M(\mathbf{x}) \leqslant -\rho(\epsilon) \quad \text{and} \quad \|\mathbf{D}(\mathbf{x})\| \geqslant \rho(\epsilon).$$

Then the sequence $\{\mathbf{X}_n\}$ defined by (1) converges with probability one to zero.

Proof. Expanding $-M(\mathbf{X}_{n+1})$ we obtain, with $0 \leqslant \theta \leqslant 1$,

$$-M(\mathbf{X}_{n+1}) = -M(\mathbf{X}_n) - \frac{a_n}{c_n} \langle \mathbf{D}_n, \mathbf{Y}_n \rangle - \frac{a_n^2}{2c_n^2} \left\langle Y_n, \mathbf{A}\left(\mathbf{X}_n + \theta \frac{a_n}{c_n} \mathbf{Y}_n\right) Y_n \right\rangle.$$

Taking conditional expectation for given X_n, we have

$$E\{-M(\mathbf{X}_{n+1}) | \mathbf{X}_n\} = -M(\mathbf{X}_n) - \frac{a_n}{c_n} \langle \mathbf{D}_n, \mathbf{\Delta}_n \rangle - \frac{a_n^2}{2c_n^2} E\left\{\left\langle Y_n, \mathbf{A}\left(\mathbf{X}_n + \theta \frac{a_n}{c_n} Y_n\right) \mathbf{Y}_n \right\rangle | \mathbf{X}_n \right\} \text{ a.s.}$$

Since $A(\mathbf{x})$ is a bounded matrix and σ_x^2 is bounded, we have

$$\left|E\left\{\left\langle \mathbf{Y}_n, \mathbf{A}\left(\mathbf{X}_n+\theta\frac{a_n}{c_n}\mathbf{Y}_n\right)\mathbf{Y}_n\right\rangle \middle| \mathbf{X}_n\right\}\right| < k_1\|\boldsymbol{\Delta}_n\|^2+k_2,$$

where k_1 and k_2 are suitably chosen positive constants. By the hypothesis we obtain

$$\Delta_n^i = c_n\langle \mathbf{D}_n, \mathbf{u}_i\rangle + \tfrac{1}{2}c_n^2\langle u_i, A(\mathbf{x}_n+\theta^i c_n\mathbf{u}_i)\,\mathbf{u}_i\rangle \quad (i = 1, \ldots, k),$$

where $\Delta_n^{(i)}$ is the ith component of $\boldsymbol{\Delta}_n$ and $0 \leqslant \theta^i \leqslant 1$ for $i = 1, \ldots, k$. Hence

$$\langle \mathbf{D}_n, \boldsymbol{\Delta}_n\rangle = c_n\|\mathbf{D}_n\|^2+\tfrac{1}{2}c_n^2\langle \mathbf{D}_n, \bar{\mathbf{A}}_n\rangle,$$

$$\|\boldsymbol{\Delta}_n\|^2 = c_n^2\|\mathbf{D}_n\|^2+c_n^3\langle \mathbf{D}_n, \bar{\mathbf{A}}_n\rangle+\tfrac{1}{4}c_n^4\|\bar{\mathbf{A}}_n\|^2.$$

Now by hypothesis, $\|\bar{\mathbf{A}}_n\|$ is bounded, say $\|\bar{\mathbf{A}}_n\|^2 \leqslant k_3$. Then

$$\|\langle \mathbf{D}_n, \mathbf{A}_n\rangle\|^2 \leqslant k_3\|\mathbf{D}_n\|^2.$$

After some computation we find

$$E\{-M(\mathbf{X}_{n+1})/\mathbf{X}_n\} \leqslant -M(\mathbf{X}_n)-a_n\{\|\mathbf{D}_n\|^2[1-\tfrac{1}{2}k_1a_n]$$
$$-\|\mathbf{D}_n\|k_3^{\frac{1}{2}}[\tfrac{1}{2}c_n-\tfrac{1}{2}k_1a_nc_n]\}+\tfrac{1}{3}k_1k_2a_n^2c_n^2+\tfrac{1}{2}k_2a_n^2/c_n^2 \text{ a.s.},$$

where n is chosen so large that $[1-\frac{1}{2}k_1a_n]$ and $[c_n-k_1a_nc_n]$ are both non-negative.

Let λ_n be a sequence of random variables defined by

$$\lambda_n = \begin{cases} 1 & \text{if } \|\mathbf{D}_n\| \geqslant 1, \\ 0 & \text{otherwise.}\end{cases}$$

We note that for n sufficiently large we have

$$a_n\{\|D_n\|^2[1-\tfrac{1}{2}k_1a_n]-\lambda_n\|\mathbf{D}_n\|k_3^{\frac{1}{2}}[\tfrac{1}{2}c_n-\tfrac{1}{2}k_1a_nc_n]\} \geqslant 0. \qquad (2)$$

Hence, for such n we obtain

$$E\{-M(\mathbf{X}_{n+1})|\mathbf{X}_n\} \leqslant -M(\mathbf{X}_n)+a_nc_n(1-\lambda_n)\|\mathbf{D}_n\|k_3^{\frac{1}{2}}|\tfrac{1}{2}-\tfrac{1}{2}k_1a_n|$$
$$+\tfrac{1}{8}k_1k_3a_n^2c_n^2+\tfrac{1}{2}k_2a_n^2/c_n^2 \text{ a.s.}$$

This inequality is still preserved if we take conditional expectation with respect to $M(\mathbf{X}_n)$ on both sides. But now we note that

$$\sum_{j=1}^{n} a_jc_jk_3^{\frac{1}{2}}[\tfrac{1}{2}-\tfrac{1}{2}k_1a_n]E\{(1-\lambda_n)\|\mathbf{D}_n\||M(\mathbf{X}_n)\}$$

converges a.s. and that

$$\sum_{1}^{n} \tfrac{1}{8} k_1 k_3 a_j^2 c_j^2 \quad \text{and} \quad \sum_{1}^{n} \tfrac{1}{2} k_2 a_j^2 / c_j^2 \quad \text{both converge.}$$

These follow from the conditions assumed on $\{a_n\}$ and $\{c_n\}$ and the definition of λ_n. Hence, we may apply the result of Problems [Appendix II, §4] to obtain that $M(\mathbf{X}_n)$ converges a.s. to a random variable. Now we note that $\sum_{1}^{n} a_j$ diverges to $+\infty$ and that $M(\mathbf{X}_n) \leqslant 0$. Hence the series

$$\sum_{j=1}^{n} a_j E\{\|\mathbf{D}_j\|^2 [1 - \tfrac{1}{2} k_1 a_j] - \lambda_j \|\mathbf{D}_j\| k_3^{\frac{1}{2}} [\tfrac{1}{2} c_j - \tfrac{1}{2} k_1 a_j c_j]\}$$

converges. This, together with (2) insures the existence of a subsequence $\mathbf{D}_{n_r}$ with the property $P[\lim_{n \to \infty} \mathbf{D}_{n_r} = 0] = 1$. Hence $\mathbf{X}_{n_r}$ converges a.s. to zero. Since $M(\mathbf{x})$ is continuous and

$$M(\mathbf{0}) = \mathbf{0}, \text{ we have } \quad P\{\lim_{r \to \infty} M(\mathbf{X}_n) = 0\} = 1,$$

which implies the desired result. One can find application of this method in Gray [42] to the problem of random circuits.

5. Problems

(1) Let R^p denote the p-dimensional Euclidean space ($p < \infty$). Define the sequence $\{\mathbf{X}_n\}$ of R^p-valued random variables by

$$\mathbf{X}_{n+1} = \mathbf{X}_n - D_n A^{-1} (\mathbf{Y}_n - \boldsymbol{\alpha}_0), \tag{1}$$

where $\mathbf{Y}_n$ is an R^p-valued random variable with conditional distribution given $\mathbf{X}_1, \ldots, \mathbf{X}_n$ the same as the distribution of $Y(\mathbf{x}_n)$; where A is an invertible linear transformation of R^p into itself such that

$$\mathbf{D}_n = d_n \mathscr{I} + J_n, \tag{2}$$

where $d_n \geqslant 0$ and such that

$$\sum_{1}^{\infty} d_n = \infty, \quad \sum_{1}^{\infty} d_n^2 < \infty \tag{3}$$

where $\mathscr{J}$ denotes the identity transformation of R^p into itself and where J_n is a linear transformation of R^p into itself such that

$$\|d_n^{-1} J_n\| \to 0 \quad \text{as} \quad n \to \infty. \tag{4}$$

Let $(\Omega, \mathscr{A}, P)$ be a probability measure space, and for each $\mathbf{X} \in R^p$, $Y(\mathbf{x})$ be an R^p-valued random variable on $(\Omega, \mathscr{A}, P)$ such that $E\| Y(\mathbf{x})\| < \infty$. Then $E[Y(\mathbf{x})]$ exists and let $M(\mathbf{x}) = E(Y(\mathbf{x}))$ be Borel measurable. (5)

There exists a non-empty class $\mathbf{A}$ of invertible linear transformations of R^p into R^p, such that for each $A \in \mathbf{A}$,

(i) for some $\boldsymbol{\theta} \in R^p$ and for all $\epsilon > 0$,

$$\inf_{\epsilon \leqslant \|\mathbf{x}-\boldsymbol{\theta}\| < 1/\epsilon} \langle \mathbf{x} - \boldsymbol{\theta}, A^{-1}[M(\mathbf{x}) - \alpha_0] \rangle > 0.$$

(ii) For constant $k > 0$ and for all $\mathbf{x} \in R^p$,

$$\|M(\mathbf{x}) - \boldsymbol{\alpha}_0\| \leqslant k\|\mathbf{x} - \boldsymbol{\theta}\|$$

with $\boldsymbol{\theta}$ as in (i),

(iii) $\sup_{\mathbf{x} \in R^p} E\| Y(\mathbf{x}) - M(\mathbf{x})\|^2 < \infty$.

Prove that, if $A \in \mathbf{A}$ $\mathbf{X}_n \to \boldsymbol{\theta}$ with probability one and

$$E\|\mathbf{X}_n - \boldsymbol{\theta}\|^2 \to 0 \quad \text{as} \quad n \to \infty.$$

(2) Assume conditions of (1) and in addition that for some $A_0 \in \mathbf{A}$, we have

$$M(\mathbf{x}) = \alpha_0 + A_0(\mathbf{x} - \boldsymbol{\theta}) + \delta(\mathbf{x})$$

with $\boldsymbol{\theta}$ as in (i) of Problem (1) and $\delta(\mathbf{x})$ be such that

$$\|\delta(\mathbf{x})\| = 0\ (\|\mathbf{x} - \boldsymbol{\theta}\|^2) \quad \text{as} \quad \|\mathbf{x} - \boldsymbol{\theta}\| \to 0.$$

Choose A in such a way that $A \in \mathbf{A}$ and $A^{-1}A_0 \geqslant \frac{1}{2}\mathscr{J}$ and $d_n = 1/n$ then prove that $n^\lambda(\mathbf{x}_n - \boldsymbol{\theta}) \to 0$ with probability one as $n \to \infty$ where $0 \leqslant \lambda < \frac{1}{2}$. (Venter)

(3) Let $\{\mathbf{u}_1, \ldots, \mathbf{n}_p\}$ be orthogonal basis in R^p. Let $\{c_n\}$ and $\{d_n\}$ be sequences of positive numbers such that

$$c_n \to 0, \quad \sum_1^\infty d_n = \infty \quad \text{and} \quad \sum_1^\infty (d_n/c_n)^2 < \infty. \tag{1}$$

Let $\{\mathbf{Y}_n^{(i)}, \mathbf{Y}_n^{-i};\ 1 \leqslant i \leqslant p,\ n \geqslant 1\}$ be a family of random variables such that, conditional upon

$$\{\mathbf{Y}_n^{(i)}, \mathbf{Y}_m^{(-i)};\ 1 \leqslant i \leqslant p,\ 1 \leqslant m \leqslant n-1\}$$

the random variables $\{\mathbf{Y}_n^{(i)}, \mathbf{Y}_n^{(-i)}; 1 \leqslant i \leqslant p\}$ are jointly independently distributed according to the distributions of

$$\{Y(\mathbf{X}_n + c_n U_i),\ Y(X_n - c_n U_i),\ 1 \leqslant i \leqslant p\},$$

where the sequence $\{\mathbf{X}_n\}$ of random variables is generated by

$$\mathbf{X}_{n+1} = \mathbf{X}_n - D_n B^{-1} \Delta_{c_n} \mathbf{Y}_n \tag{2}$$

and where $\{\mathbf{D}_n\}$ is a sequence of linear transformations of R^p into itself such that

$$\|D_n - d_n \mathscr{I}\| \to 0 \quad \text{as} \quad n \to \infty, \tag{3}$$

B is a self-adjoint linear transformation of R^p into itself; and

$$\left.\begin{aligned} \Delta_{c_n} Y_n &= \sum_{i=1}^{p} \frac{Y_n^{(-i)} - Y_n^{(i)}}{2c_n} u_i, \\ \Delta_{c_n} M(x) &= \sum_{i=1}^{p} \frac{M(x - cu_i) - M(x + cu_i)}{2c} u_i. \end{aligned}\right\} \tag{4}$$

Letting

$$\left.\begin{aligned} Z_n^{(i)} &= Y_n^{(i)} - M(\mathbf{X}_n + C_n U_i), \\ Z_n^{(-i)} &= Y_n^{(-i)} - M(\mathbf{X}_n - C_n U_i) \end{aligned}\right\} \tag{5}$$

and

$$\Delta_{c_n} Z_n = \Delta_{c_n} Y_n - \Delta_{c_n} M(X_n) \tag{6}$$

so that

$$E[\Delta_{c_n} \mathbf{Z}_n | \mathbf{X}_1, \ldots, \mathbf{X}_n] = 0 \text{ a.s.}$$

(i) There exists a non-empty class B of self-adjoint invertible linear transformations of R^p into itself, such that, for some element $\boldsymbol{\theta} \in R^p$, for each $B \in \mathrm{B}$ and for each $\epsilon > 0$, there exists a number $C_{\epsilon, B}$ with $0 < C_{\epsilon, B} < \epsilon$ such that

$$\inf(\mathbf{X} - \boldsymbol{\theta}, B^{-1} \Delta_c M(x)) > 0, \quad \epsilon \leqslant \|x - \theta\| \leqslant 1/\epsilon, \quad 0 < c < C_{\epsilon, B}.$$

(ii) For some $c_1 > 0$, for all c such that $0 < c < c_1$, and for each $\mathbf{x} \in R^p$,

$$|M(\mathbf{x} + cu_i) - M(\mathbf{x})| \leqslant k_1 + k_2 \|\mathbf{x}\| \quad (i = 1, \ldots, p),$$

where k_1 and k_2 are positive constants.

(iii) $\sup_{\mathbf{x} \in R^p} E|Z(\mathbf{x})|^2 < \infty$ and if $B \in \mathrm{B}$ and

$$\mathbf{X}_{n+1} = \mathbf{X}_n - D_n B^{-1} \Delta_{c_n} M(X_n) - D_n B^{-1} \Delta_{c_n} \mathbf{Z}_n.$$

Then prove that $X_n \to \theta$ with probability one and

$$E\|\mathbf{x}_n - \boldsymbol{\theta}\|^2 \to 0 \quad \text{as} \quad n \to \infty.$$

(4) Assume conditions of Problem 3.

Define two stage multivariate stochastic approximation and state conditions under which procedure has desired convergence property. (Venter)

CHAPTER 6

ASYMPTOTIC NORMALITY

1. Introduction

In Chapters 2, 3, and 5 there was a discussion of random variables generated by sequential schemes of stochastic approximation. In this chapter conditions will be discussed where the random variable is asymptotically normal Two main approaches are available. The first method is due to Chung [12] who by the classical method of moments shows asymptotic normality. This method is adapted by Burkholder [10] to prove Theorem 1 and gives a good feeling for the structure of a variate. The second is an elegant method of Sacks [67] who, by extending a result on characteristic functions stated in Loève [58] to the multivariate case, proves the asymptotic normality. Fabian [53] has very recently given another proof of asymptotic normality which is the subject of Theorem 4. Hodges and Lehmann [45] prove a result of asymptotic normality which is easily employed in practice.

A volume could be written on applications of normality but obviously such an attempt cannot be made here. We confine ourselves to a discussion of confidence intervals and of problems which throw light on this utility. Schalkwijk [68] gives a further application.

2. Method of moments

THEOREM 1. (Burkholder). Suppose $\{X_n\}$ is a stochastic approximation process of the type A_0, $\{v_n\}$ is a real number sequence, $\{c_n\}$ is a positive number sequence, θ is a real number and each of $\beta, I, \overline{T}, \mathrm{v}, \bar{v}, \sigma^2, \gamma, \xi, c$, and d is a positive number such that

(i) $R_n(v_n) = 0$ for all n in the set N of positive integers.

(ii) The function sequence $\{T_n\}$, where for each n in N,

$$T_n(x) = \frac{R_n(x)}{c_n(x - v_n)} \quad \text{if} \quad x \neq v_n$$
$$= \beta \quad \text{if} \quad x = v_n$$

is continuously convergent at θ to β (that is, $\{x_n\}$ is sequence of real number with limit θ, then $T_n(x_n) \to \beta$ as $n \to \infty$) and satisfies $\mathbf{T} \leqslant T_n(x) \leqslant \overline{T}$ for all n, x in $N \times R$.

(iii) $\{V_n\}$ is continuously convergent at θ to σ^2 and satisfies $\mathbf{V} \leqslant V_n(x) \leqslant \overline{V}$ for all n, x in $N \times R$.

(iv) If r is in N then $\sup_{n,x} E[Z_n(x) - R_n(x)]^r < \infty$.

(v) $G_n(y, .)$ is Borel measurable for each y in R.

(vi) $v_n - \theta = O(n^{-\gamma}), \xi < \frac{1}{2}, \quad (\xi < \gamma), na_n c_n \to d > \xi/I,$

$$n^{\xi+\frac{1}{2}} a_n \to c \quad \text{as} \quad n \to \infty.$$

(vii) All moments of X_i are finite.

Then if r is in N

$$\lim_{n \leftarrow \infty} n^{r\xi} E(X_n - \theta)^r = \left[\frac{\sigma^2 c^2}{2\beta d - 2\xi}\right]^{\frac{1}{2}r} (r-1)(r-3)\ldots 3\cdot 1$$

$$\text{if } r \text{ is even,}$$

$$= 0 \quad \text{if } r \text{ is odd,}$$

which implies that $n^{\xi}(X_n - \theta)$ is asymptotically normal with mean zero and variance $\sigma^2 c^2/2(\beta d - \xi)$.

Proof. If n is in N let $\xi_n^{(r)} = E(X_n - \theta)^r$, if r is a non-negative integer. $\beta_n^r = E|X_n - \theta|^r$, if $r \geqslant 0$ and $\xi_n = \xi_n^{(2)}$. It can be easily verified that these expectations exist under the conditions of the theorem. If r is in N then

$$(X_{n+1} - \theta)^r = (X_n - \theta - a_n z_n)^r = \sum_{k=0}^{r} \binom{r}{k} (a_n)^k (X_n - \theta)^{r-k} Z_n^k.$$

Let $H_k(r, n) = E[(X_n - \theta)^{r-k} Z_n^k]$ for each k, r, n in N such that $k \leqslant r$ we have that

$$\xi_{n+1}^{(1)} = \xi_n^{(1)} - a_n H_1(1, n), \tag{1}$$

$$\xi_{n+1} = \xi_n - 2a_n H_1(2, n) + a_n^2 H(2, n) \tag{2}$$

and if $r \geqslant 2$ then

$$\xi_{n+1}^{(r)} = \xi_n^{(r)} - ra_n H_1(r, n) + \binom{r}{2} a_n^2 H(r, n) + \sum_{k=3}^{r} \binom{r}{k} (-a_n)^k H_k(r, n). \tag{3}$$

By conditions (i) and (ii)

$$R_n(x) = C_n T_n(x)(x - v_n) \quad \text{for} \quad n, x,$$

$$\begin{aligned} H_1(r, n) &= E\{E(X_n - \theta)^{r-1} Z_n | x_n\} \\ &= E[(X_n - \theta)^{r-1} R_n(x_n)] \\ &= C_n E[T_n(x_n)(X_n - \theta)^r] \\ &\quad + C_n(\theta - v_n) E[T_n(x_n)(X_n - \theta)^{r-1}]. \end{aligned} \tag{4}$$

Thus, if r is even then

$$H_1(r, n) \geqslant C_n \mathbf{T} \xi_n^{(r)} - C_n |\theta - v_n| \overline{T} \beta_n^{(r-1)}. \tag{5}$$

Since by lemma 17 of Appendix 3 $[\beta_n^r]^{1/r}$ is non-decreasing in r for $r > 0$, (5) implies that if r is even then

$$H_1(r, n) \geqslant C_n \mathbf{T} \xi_n^{(r)} - C_n |\theta - v_n| \overline{T} [\xi_n^{(r)}]^{(r-1)/r}. \tag{6}$$

Also, since $2\beta_n^{(r-1)} \leqslant \beta_n^{(r-2)} + \beta_n^{(r)}$ for $r \geqslant 2$, we have by (5) and relation $|\theta - v_n| = o(1)$ which is implied by (vi), that if r is even then

$$H_1(r, n) \geqslant [C_n \mathbf{T} + o(C_n)] \xi_n^{(r)} - C_n |\theta - v_n| \overline{T} \frac{\xi_n^{(r-2)}}{2}. \tag{7}$$

If r is in N and $r \geqslant 2$ then

$$\begin{aligned} H_2(r, n) &= E[(X_n - \theta)^{r-2} E\{Z_n - R_n(X_n) + R_n(X_n)\}^2 | X_n] \\ &= E[V_n(X_n)(X_n - \theta)^{r-2}] \\ &\quad + C_n^2 E[T_n^2(X_n)(X_n - \theta)^{r-2} \times (X_n - v_n)^2]. \end{aligned}$$

Since

$$(X_n - v_n)^2 = (X_n - \theta + \theta - v_n)^2 \leqslant 2|X_n - \theta|^2 + 2|\theta - v_n|^2$$

we obtain that

$$|H_2(r, n)| \leqslant [\overline{V} + o(1)] \beta_n^{(r-2)} + 2C_n^2 \overline{T}^2 \beta_n^{(r)}. \tag{8}$$

If each of r and k is in N and $r \geqslant k$ then

$$\begin{aligned} |H_k(r, n)| &\leqslant E[|X_n - \theta|^{r-k} E\{|Z_n - R_n(X_n) + R_n(X_n)|^k | X_n\}] \\ &\leqslant 2^k E[|X_n - \theta|^{r-k} E\{|Z_n - R_n(X_n)|^k + |R_n(X_n)|^k | X_n\}] \end{aligned}$$

and it follows that

$$|H_k(r, n)| = O(1) \beta_n^{(r-k)} + O(1) c_n^k \beta_n^{(r)}. \tag{9}$$

(I) We shall now prove that if r is a positive number then there is a real number $\bar{\beta}_r$ such that

$$\limsup_{n\to\infty} n^{r\xi}\beta_n^{(r)} \leqslant \bar{\beta}_r. \tag{10}$$

Since $[\beta_n^r]^{1/r}$ is non-decreasing in r for $r > 0$ it suffices to show that (10) holds for each even natural number r. Let us consider the case $r = 2$, using (7) and (8) in (2) gives

$$\xi_{n+1} \leqslant \xi_n \left[1 - \frac{2na_n C_n T + o(1)}{n}\right] + \frac{n^{2\xi+1}a_n^2 \bar{V} + o(1)}{n^{2\xi+1}} + \frac{O(1)}{n^{\gamma+1}}$$

using (6) and (8) in (2) gives

$$\xi_{n+1} \leqslant \xi_n \left[1 - \frac{2na_n C_n \mathrm{T} + o(1)}{n}\right] + \frac{n^{2\xi+1}a_n^2 \bar{V} + o(1)}{n^{2\xi+1}} + O(1)\, \xi_n^{\frac{1}{2}}/n^{\gamma+1}.$$

The assumption of the theorem and the above relations imply, by lemma 4 of Appendix 3, that

$$\limsup_{n\to\infty} n^{2\xi}\, \xi_n \leqslant c^2\, \bar{V}/2(Id - \xi).$$

Thus (10) holds for $r = 2$.

Suppose that r is even $r > 2$ and that (10) holds for each even natural number less than r. Then, of course, $\beta_n^{(k)} = O(n^{-k\xi})$ for each positive number $k \leqslant r-2$. Using this fact and (9) gives

$$\sum_{k=3}^{r} \binom{r}{k} (-a_n)^k H_k(r,n) = o(n^{-r\xi-1}) + o(n^{-1})\, \xi_n^{(r)}. \tag{11}$$

Substituting (7), (8) and (11) in (3) gives

$$\xi_{n+1}^{(r)} \leqslant \xi_{(n)}^{(r)} \left[1 - \frac{r dI + o(1)}{n}\right] + \frac{\binom{r}{2} c^2 \bar{V}\bar{\beta}_{r-2} + o(1)}{n^{r\xi+1}} + \frac{O((1)}{n^{\gamma+1}}.$$

Substituting (6), (8) and (11) in (3) gives

$$\xi_{n+1}^{(r)} \leqslant \xi_n^{(r)} \left[1 - \frac{r dI + o(1)}{n}\right] + \frac{\binom{r}{2} c^2\, \bar{V}\bar{\beta}_{r-2} + o(1)}{n^{r\xi+1}} + \frac{O(1)[\xi_n^{(r)}]^{(r-1)r}}{n^{\gamma+1}}.$$

Thus by lemma 4 of Appendix 3

$$\limsup_{n\to\infty} n^{r\xi}\, \xi_n^r \leqslant (r-1)\, c^2 V \bar{\beta}_{r-2/2(Id-\xi)}.$$

By induction, (10) holds for each even natural number. Thus, (I) is proved.

(II) Next it will be shown that if $\delta > 0$ and r is in N then

$$E\{|X_n - \theta|^r \,|\, |X_n - \theta| \geqslant \delta\} \times P\{|X_n - \theta| \geqslant \delta\} = o(1)\,\beta_n^{(r)}.$$

If $r \geqslant 0$, $\delta > 0$, $q > 0$ then by (I)

$$\begin{aligned} &E\{|X_n - \theta|^r / |X_n - \theta| \geqslant \delta\} \times P\{|X_n - \theta| \geqslant \delta\} \\ &\quad \leqslant \delta^{-2q/\xi} E\{|X_n - \theta|^{r+2q/\xi}| \, |X_n - \theta| \geqslant \delta\} \times P\{|X_n - \theta| \geqslant \delta\} \\ &\quad \leqslant \delta^{-2q/\xi} \times \beta_n^{(r+2q/\xi)} = o(n^{-q}). \end{aligned} \tag{12}$$

Since $X_{n+1} - E(X_{n+1}|X_n) = -a_n[Z_n - R_n(X_n)]$ we have that

$$\begin{aligned} E\{[X_{n+1} - \theta]^2 | X_n\} &\geqslant E\{[X_{n+1} - E(X_{n+1}|X_n)]^2 | X_n\} \\ &= a_n^2 E([Z_n - R_n(X_n)]^2 | X_n) \\ &= a_n^2 V_n(X_n) \geqslant a_n^2 \mathrm{V}. \end{aligned}$$

Thus

$$(n+1)^{2\lambda}\xi_{n+1} \geqslant n^{2\lambda}\xi_{n+1} \geqslant n^{2\lambda}a_n^2\,\mathrm{V} = c^2\mathrm{V} + o(1),$$

where $\lambda = \xi + \frac{1}{2}$, therefore if $r \geqslant 2$ then

$$\liminf_{n\to\infty} n^{2\lambda}\beta_n^{(r)} \geqslant \liminf_{n\to\infty} (n^{2\lambda}\xi_n)^{r/2} \geqslant (c^2\mathrm{V})^{r/2}. \tag{13}$$

Furthermore

$$\begin{aligned} n^{2\lambda}\beta_n^{(1)} &\geqslant n^{2\lambda}E\{|x_n - \theta| \,|\, |x_n - \theta| < 1\} P\{|x_n - \theta| < 1\} \\ &\geqslant n^{2\lambda}E\{|x_n - \theta|^2 / |x_n - \theta| < 1\} P\{|x_n - \theta| < 1\} \\ &= n^{2\lambda}\xi_n + o(1) \quad \text{using (12)} \end{aligned}$$

$$\liminf_{n\to\infty} n^{2\lambda}\beta_n^{(1)} \geqslant c^2\,\mathrm{V}. \tag{14}$$

Relations (12), (13) and (14) imply the assertion (II). Using (ii), (vi), (I) and (II) in (4) gives by lemma 2 of Appendix II §2 that if r is in N then

$$\begin{aligned} H_1(r, n) &= c_n \beta\xi_n^r + c_n o(1)\,\beta_n^{(r)} + c_n o(n^{-\gamma})\,\beta_n^{(r-1)} \\ &= c_n \beta\xi_n^{(r)} + c_n o(n^{-r\xi}). \end{aligned} \tag{15}$$

Similarly, if r is in N and $r \geqslant 2$ then

$$H_2(r, n) = \sigma^2\xi_n^{(r-2)} + o(n^{-(r-2)\xi}). \tag{16}$$

Also $$\sum_{k=3}^{r} \binom{r}{k} (-a_n)^k H_k(r,n) = o(n^{-r\xi-1}). \tag{17}$$

It will now be shown by induction that

$$\lim_{n\to\infty} n^{r\xi} \xi_n^{(r)} = \beta_r \quad \text{for each } r \text{ in } N, \tag{18}$$

where $$\beta_r = \left[\frac{\sigma^2 c^2}{2\beta d - 2\xi}\right]^{r/2} \quad (r-1)(r-3)(r-5)\ldots 3\cdot 1$$

$$\text{if } r \text{ is even,}$$

$$= 0 \qquad \text{if } r \text{ is odd.}$$

Consider the case $r = 1$. Using (15) in (1) gives

$$\xi_{n+1}^{(1)} = \xi_n^{(1)} \left(1 - \frac{n a_n c_n \beta}{n}\right) + o(n^{-\xi-1})$$

implying that there is a natural number n_1 such that if $n > n_1$ then

$$|\xi_{n+1}^{(1)}| \leqslant |\xi_n^{(1)}| \left(1 - \frac{n a_n c_n \beta}{n}\right) + o(n^{-\xi-1}).$$

By lemma 2, Appendix 3 using the fact that $\beta > \mathbf{T}$ we have that $\limsup_{n\to\infty} n^{\xi}|\xi_n^{(1)}| = 0$ which implies (18) holds for $r = 1$.

Consider the case $r = 2$. Using (15) and (16) in (2) gives

$$\xi_{n+1} = \xi_n \left[1 - \frac{2 n a_n c_n \beta}{n}\right] + \frac{n^{2\xi+1} a_n^2 \sigma^2 + o(1)}{n^{2\xi+1}}.$$

Thus by lemma 4, Appendix 3 (18) holds for $r = 2$.

Suppose r is a natural number greater than 2 and (18) holds for $r = 2$. Using (15), (16), (17) and induction hypothesis in (14) gives

$$\xi_{n+1}^{(r)} = \xi_n^{(r)} \left[1 - \frac{r\beta d + o(1)}{n}\right] + \frac{\binom{r}{2} \sigma^2 c^2 \beta_{r-2} + o(1)}{n^{r\xi+1}}.$$

If r is odd, we proceed as in the case $r = 1$ and find by lemma 4, Appendix 3 that $\limsup_{n\to\infty} n^r |\xi_n^{(r)}| = 0$. If r is even, lemma 4, Appendix 3 implies that

$$\lim_{n\to\infty} n^{r\xi} \xi_r^{(n)} = (r-1)\,\sigma^2 c^2 \beta_{r-2/2(\beta d - \xi)}.$$

Thus by induction, (18) holds for each natural number r.

This completes the proof of the theorem.

3. Method of characteristic function

Sacks [67] used method of characteristic function and proved asymptotic normality.

THEOREM 2 (Sacks). *Let $\{a_n\}$ be sequence of positive numbers such that*

$$\sum_1^\infty a_n = \infty, \quad \sum_1^\infty a_n^2 < \infty. \tag{1}$$

Let X_i be a fixed number of arbitrary random variables with

$$E(X_1^2) < \infty \quad \textit{and} \quad \{X_n\},$$

the sequence is defined by the recursion

$$X_{n+1} = X_n - a_n(y(X_n) - \alpha), \tag{2}$$

where $y(X_n)$ is a random variable whose conditional distribution given $x_1, x_2, \ldots, x_n$ is the same as the distribution of $y(x_n)$. Let $Z(x) = y(x) - M(x)$, where $M(x) = E(y(x))$ and assume that $M(x) = \alpha$ has a unique solution for each real value α.

Thus (2) becomes

$$X_{n+1} = X_n - a_n[M(X_n) - \alpha + Z(X_n)]. \tag{3}$$

$E(Z(x)) = 0$ for all x, and the conditional distribution of $Z(x_n)$ given $x_1, x_2, \ldots, x_n$ is the same as the distribution of $Z(x_n)$. We note that as a consequence of this

$$E(Z(x_n) | Z(x_1), \ldots, Z(x_{n-1})) = 0 \tag{4}$$

with probability one, and make the following assumptions:

(i) M is a Borel measurable function; $M(\theta) = \alpha$ and

$$(x-\theta)(M(x)-\alpha) > 0 \quad \text{for all } x \neq 0.$$

(ii) For some positive constants k and k_1, and for all x

$$k|x-\theta| \leqslant |M(x)-\alpha| \leqslant k_1|x-\theta|.$$

(iii) For all x

$$M(x) = \alpha + \alpha_1(x-\theta) + \delta(x, \theta),$$

where $\delta(x, \theta) = o(|x-\theta|)$ as $x - \theta \to 0$ and where $\alpha_1 > 0$.

(iv) (a) $\sup_x E(Z^2(x)) < \infty$; (b) $\lim_{x\to\theta} E(Z^2(x)) = \sigma^2$.

(v) $\lim_{R\to\infty} \lim_{\epsilon\to 0^+} \sup_{|x-\theta|<} \int_{\{\epsilon|Z(x)|>R\}} Z^2(x)\,dp = 0.$

Let $a_n = An^{-1}$ for $n > 0$, where A is such that $2kA > 1$. Then $n^{\frac{1}{2}}(X_n - \theta)$ is asymptotically normally distributed with mean zero and variance $A^2\sigma^2(2A\alpha_1 - 1)^{-1}$.

Proof. Without loss of generality let $\alpha = \theta = 0$ and abbreviate $\delta(X_n, 0)$, $M(X_n)$ and $Z(X_n)$ by δ_n, M_n and Z_n respectively. Using (iii) and (3) we have

$$X_{n+1} = (1 - A\alpha_1 n^{-1}) X_n - An^{-1}\delta_n - An^{-1}Z_n. \tag{5}$$

Let $a = A\alpha_1$ and let $\beta_{mn} = \prod_{j=m+1}^{n} (1-a_j)$ for $0 \leqslant m \leqslant n$ and $\beta_{mn} = 1$ for $m = n$, where $a_j = aj^{-1}$. Iteration of (5) yields

$$X_{n+1} = \beta_{0n} X_1 - A \sum_{k=1}^{n} k^{-1}\beta_{kn}\delta_k - A \sum_{m=1}^{n} k^{-1}\beta_k nZ_k. \tag{6}$$

Let
$$h_n = \left(\sum_{m=1}^{n} a^2k^{-2}\beta_{kn}^2\right)^{\frac{1}{2}}.$$

Then by lemma 15, Appendix 3

$$h_n \sim (2a-1)^{\frac{1}{2}} a^{-1} n^{\frac{1}{2}}.$$

Hence, proving that $n^{\frac{1}{2}}X_n$ is asymptotically normal with mean 0 and variance $A^2\sigma^2(2a-1)^{-1}$ is equivalent to proving h_nX_n is asymptotically normal with mean 0 and variance $A^2\sigma^2a^{-2}$. We have to show that

$$h_n\beta_{0n} \to 0 \quad \text{as} \quad n \to \infty, \tag{7}$$

$$h_n \sum_{m=1}^{n} a_k^{-1}\beta_{kn}\delta_k \to 0 \quad \text{as} \quad n \to \infty, \tag{8}$$

$$h_n \sum_{m=1}^{n} ak^{-1}\beta_{kn}Z_k \text{ is asymptotically normal } (0, \sigma^2). \tag{9}$$

Let $c_k = 1$ in lemma 12, Appendix 3, for all k, it follows from the lemma that (7) holds. To prove (9) we will invoke Theorem 6, Appendix 2 with $q = 1$ and $U_{nk} = h_n ak^{-1}\beta_{kn}Z_k$. To see that we can do so, observe first that by (4)

$$E(U_{nk}|U_{n1}, \ldots, U_{n,k-1}) = E(U_{kn}|Z_1, \ldots, Z_{k-1}) = 0.$$

Let $\phi_{nk} = 1$ if $|U_{nk}| \geqslant \epsilon$ and $\phi_{nk} = 0$ otherwise and observe that in order to verify condition of the Theorem 6, Appendix 2 we have to check that

$$\sum_{k=1}^{n} E(\phi_{nk} U_{nk}^2) \to 0 \quad \text{as} \quad n \to \infty,$$

what is the same

$$h_n^2 \sum_{k=1}^{n} a^2 k^{-2} \beta_{kn}^2 E(\phi_{nk} Z_k^2) \to 0 \quad \text{as} \quad n \to \infty. \tag{10}$$

Noticing by lemma 12, Appendix 3 and Problem 1, Appendix 3 that $\phi_{nk} = 1$ implies, for some $\epsilon' > 0$, that

$$|Z_k| \geqslant \epsilon' n^{a-1/2} k^{1-a} > \epsilon^1 k^{\frac{1}{2}}.$$

Applying (v) one can obtain

$$\lim_{k\to\infty} E(\phi_k' Z_k^2) = 0,$$

where

$$\phi_k' = 1 \quad \text{if} \quad |Z_k| > \epsilon'^{k^{\frac{1}{2}}}$$

$$= 0 \quad \text{otherwise.} \tag{11}$$

Since $\phi_k' \geqslant \phi_{nk}$, applying lemma 13, Appendix 3 with $c_k = 1$ for all k and using (11) one establishes that (10) is valid. To verify (i) of Theorem 6, Appendix 2 is equivalent to showing that

$$\lim_{n\to\infty} h_n^2 \sum_{k=1}^{n} a^2 k^{-2} \beta_{kn}^2 E|E'[Z^2(X_k) - E(Z^2(X_k))]| = 0, \tag{12}$$

where E' denotes conditional expected value with conditioning being by V_{nk} of Theorem 6, Appendix 2. Using again lemma 13, Appendix 3 it follows that it is sufficient to prove

$$\lim_{n\to\infty} E|E'[Z^2(X_k) - E(Z^2(X_k))]| = 0. \tag{13}$$

But (13) it follows easily by observing that expression between the absolute value signs is uniformly bounded by (a) (iv), so that Lebesgue's theorem is applicable, and by observing that (b) (iv) together with the convergence of x_k to θ with probability one imply

$$\lim_{k\to\infty} E'[Z^2(X_k)] = \lim_{k\to\infty} E[Z^2(X_k)] = \sigma^2. \tag{14}$$

(14) and lemma 13 Appendix 3 also shows that (2) of Theorem 6, Appendix 2 is satisfied with

$$\lim_{n\to\infty} \delta_n = \sigma^2, \tag{15}$$

where
$$\delta_n = \sum_1^n \delta_{nk} = \sum_1^n E(U_{nk} U_{nk'}).$$

This completes the verification that Theorem 6, Appendix 2 is applicable and therefore establishes (9). To prove (8) we proceed as follows.

Squaring both sides of (3) and using (iv) we get

$$E(X_{n+1}^2) = E(X_n - An^{-1} M_n)^2 + O(n^{-2}). \tag{16}$$

Then, by (i) and (ii), for ϵ sufficiently small so that $2KA - \epsilon > 1$, and for n sufficiently large, say $n > n_0$,

$$\begin{aligned} E(X_{n+1}^2) &\leqslant (1 - 2kAn^{-1} + A^2 k_1^2 n^{-2}) E(X_n^2) + O(n^{-2}) \\ &\leqslant (1 - (2kA - \epsilon) n^{-1}) E(X_n^2) + D_1 n^{-2}. \end{aligned} \tag{17}$$

Let $p = 2kA - \epsilon$ and let β_{kn} be defined as earlier with $a_j = pj^{-1}$. Choose n_0 large enough so that $p < n_0$ (this is to guarantee that $\beta_{kn} > 0$ for $k \geqslant n_0$ so that (18) can hold). Iteration of (17) yields

$$\begin{aligned} E(X_{n+1}^2) &\leqslant D_1 \sum_{k=n_0+1}^{n} k^{-2} \beta_{kn} + \beta_{n_0 n} E(X_{n+1}^2) \\ &\leqslant D_2 n^{-1} + D_3 n^{-p} \\ &= O(n^{-1}), \end{aligned} \tag{18}$$

which is the estimate we require.

Let $t > 0$. Since $\delta(x) = O(|x|)$, for $t > 0$ we can find $\epsilon > 0$ with the property that

$$|\delta(x)| \leqslant t^2 |x| \quad \text{for} \quad |x| \leqslant \epsilon. \tag{19}$$

As was pointed out above $x_n \to \theta$ with probability one; hence, we can choose n_1 so that

$$P\{|X_j| \leqslant \epsilon, j \geqslant n_1\} > 1 - t. \tag{20}$$

Let $n_2 = \max[n_0, n_1]$ and such that $a < n_2 + 1$. Then denoting $h_n \sum_{k=n_2}^{n} ak^{-1} \beta_{kn}$ by V_n and $h_n \sum_{k=n_2}^{n} ak^{-1} \beta_{kn} |X_k|$ by V_n^*, and using (20),

(19) and (18), Lyapounov's inequality, and Problem 1, Appendix 3, we have for $n > n_2$

$$P[|V_n| > t] \leqslant t + P\{|V_n| > t;\ |X_j| \leqslant t, j \geqslant n_2\}$$
$$\leqslant t + P\{t^2 v_n^* > t\} \leqslant t + tEV_n^*$$
$$\leqslant t + D_4 t h_n \sum_{n_2}^{n} k^{-1} \beta_{kn} k^{-\frac{1}{2}} \leqslant D_5 t, \tag{21}$$

(where D's are appropriate constants).

(21), together with the fact that $h_n \beta_{kn} \to 0$ as $n \to \infty$ for any fixed k establishes (8) and completes the proof of the theorem.

THEOREM 3 (Sacks). *Assume the set up of the Theorem* 2 *and assumptions* (i), (iii), (iv) *and* (v) *replace* (ii) *by the following assumption* (ii)′; (ii)′ *for all x and some positive constant k_1*

$$|M(x) - \alpha| \leqslant k_1 |x - \theta|$$

and, for every t_1, t_2 such that $0 < t_1 < t_2 < \infty$

$$\inf_{t_1 \leqslant |x-\theta| \leqslant t_2} |M(x) - \alpha| > 0.$$

Let $a_n = An^{-1}$, where A is such that $A\alpha_1 > \frac{1}{2}$. Then $n^{\frac{1}{2}}(X_n - \theta)$ is asymptotically normally distributed with mean 0 *and variance* $A^2\sigma^2(2A\alpha_1 - 1)^{-1}$.

Proof. Without loss of generality let $\alpha = \theta = 0$. Let $t > 0$ be such that $A(\alpha_1 - t) > \frac{1}{2}$. Let $k = \alpha_1 - t$. Then we can find an $\epsilon > 0$ such that for $|x| \leqslant \epsilon$.

$$k|x| \leqslant |M(x)| \leqslant k_1|x|. \tag{1}$$

Define
$$M^*(x) = M(x) \quad \text{if} \quad |x| \leqslant \epsilon$$
$$= kx \quad \text{if} \quad |x| > \epsilon.$$

Since (i), (ii)′ and (iv) imply $X_n \to 0$ with probability one we can find n_0 so that for $u > 0$

$$R\{|X_j| \leqslant \epsilon, j \geqslant n_0\} > 1 - u. \tag{2}$$

Let $X_1' = x_{n_0+1}$ and define $\{X_n', n \geqslant 1\}$ by the recursion

$$X_{n+1}' = X_n' - a_{n+n_0} M^*(X_n') - a_{n+n_0} Z(X_n'). \tag{3}$$

It is clear that the assumptions of Theorem 2 together with (1) show that Theorem 6, Appendix 2 is applicable to X_n', M^*, $\{a_{n+n_0}\}$. Hence for all y

$$\lim_{n\to\infty} P\{(n_0+n)^{\frac{1}{2}} X_{n+1}' < y\} = F(y), \tag{4}$$

where F is the normal distribution functions with mean 0 and variance $A^2\sigma^2(2A\alpha_1 - 1)^{-1}$. By (2) and (4) the result of the theorem follows.

Now we discuss result of Fabian [33].

Preliminaries.

In what follows, $(\Omega, \mathfrak{F}, P)$ will be a probability space, relations between and convergence of random variables, vectors, and matrices will be meant with probability one unless specified otherwise.

We shall write $\mathbf{X}_n \sim \mathfrak{L}$ if $\mathbf{X}_n$ is asymptotically $\mathfrak{L}$-distributed and $\mathbf{X}_n \sim \mathbf{Y}_n$, for two sequences of random vectors, if for any $\mathfrak{L}$, $\mathbf{X}_n \sim \mathfrak{L}$ if and only if $\mathbf{Y}_n \sim \mathfrak{L}$.

The indicator function of a set A will be denoted by χA, the expectation and conditional expectation by E and E_p. R^k is the k-dimensional Euclidean space whose elements are considered to be column vectors, $R = R^1$, $R^{k\times k}$ is the space of all real $k \times k$ matrices. The symbols $\mathbf{R}$, $\mathbf{R}^k$ $\mathbf{R}^{k\times k}$ denote sets of all measurable transformations from $(\Omega, \mathfrak{F})$ to R, R^k, $R^{k\times k}$, respectively. The unit matrix in $R^{k\times k}$ is denoted by I and $\| \;\|$ is the Euclidean norm. With h_n a sequence of numbers, $o(h_n)$, $O(h_n)$, $o_u(h_n)$, $O_u(h_n)$ denote sequences g_n, G_n, q_n, Q_n, say, of elements in one of the sets, $\mathbf{R}$, $\mathbf{R}^k$, $\mathbf{R}^{k\times k}$ such that $h_n^{-1} g_n \to 0$, $\|h_n^{-1} G_n\| \leqslant f$ for an $f \in \mathbf{R}$ and all n, $h_n^{-1} q_n \to 0$ uniformly on a set of probability one, $\|h_n^{-1} Q_n\| \leqslant K$ for a $K \in R$ and all n. In special cases $o(h_n)$ may be constant on Ω and considered as a sequence with elements in R, R^k or $R^{k\times k}$. Similarly in other cases.

THEOREM 4. (Fabian). Suppose k is a positive integer, F_n a non-decreasing sequence of σ-fields, $F_n \subset \mathfrak{F}$; suppose

$$\mathbf{U}_n, \mathbf{V}_n, \mathbf{T}_n \in \mathbf{R}^k;\ \mathbf{T} \in R^k;\ \Gamma_n,\ \Phi_n \in \mathbf{R}^{k\times k},\ \Sigma, \Gamma, \Phi, P \in R^{k\times k};$$

Γ is positive definite; P is orthogonal and $P'\Gamma P = \Lambda$ diagonal. Suppose Γ_n, Φ_n, $\mathbf{V}_{n-1}$ are F_n-measurable, $C, \alpha, \beta, \in R$ and

$$\Gamma_n \to \Gamma, \Phi_n \to \Phi, \mathbf{T}_n \to \mathbf{T} \quad \text{or} \quad E\|\mathbf{T}_n - \mathbf{T}\| \to 0, \tag{1}$$

$$E_{F_n}\mathbf{V}_n = 0, \quad C > \|E_{F_n}\mathbf{V}_n\mathbf{V}_n' - \Sigma\| \to 0, \tag{2}$$

and, with $\sigma_{j,r}^2 = E\chi\{\|\mathbf{V}_j\|^2 \geqslant rj^\alpha\}\|\mathbf{V}_j\|^2$, let either

$$\lim_{j\to\infty} \sigma_{j,r}^2 = 0 \quad \text{for every} \quad r > 0, \tag{3}$$

or

$$\alpha = 1, \lim_{n\to\infty} \frac{1}{n}\sum_{j=1}^{n} \sigma_{j,r}^2 = 0 \quad \text{for every} \quad r > 0. \tag{4}$$

Suppose that, with $\lambda = \min_i \Lambda^{(ii)}$, $\beta_+ = \beta$ if $\alpha = 1$, $\beta_+ = 0$ if $\alpha \neq 1$,

$$0 < \alpha \leqslant 1, \quad 0 \leqslant \beta, \quad \beta_+ < 2\lambda \tag{5}$$

and

$$\mathbf{U}_{n+1} = (I - n^{-\alpha}\Gamma_n)\,\mathbf{U}_n + n^{-(\alpha+\beta)/2}\Phi_n\mathbf{V}_n + n^{-\alpha-\beta/2}\mathbf{T}_n. \tag{6}$$

Then the asymptotic distribution of $n^{\beta/2}\mathbf{U}_n$ is normal with mean $(\Gamma - (\beta_+/2)\,I)^{-1}\mathbf{T}$ and covariance matrix PMP', where

$$M^{(ij)} = (P'\Phi\Sigma\Phi'P)^{(ij)}\,(\Lambda^{(ii)} + \Lambda^{(jj)} - \beta_+)^{-1}. \tag{7}$$

Proof. Let $\eta(m, S)$ be the normal distribution with mean m and covariance matrix S, let $\eta = \eta((\Gamma - (\beta_+/2)\,I)^{-1}\mathbf{T},\ PMP')$. As the first step we shall show that we may, without loss of generality, assume that

$$\Gamma = \Lambda, \quad P = I, \quad \beta = \beta_+ = 0, \quad \|\Gamma_n - \Lambda\| = o_u(1),$$
$$\Phi = \Phi_n = I, \quad \mathbf{T}_n = \mathbf{U}_1 = 0. \tag{8}$$

If we consider $\tilde{\mathbf{U}}_n = P'\mathbf{U}_n$ then $\tilde{U}_n$ satisfies again the assumptions of the theorem, but with $\tilde{\Gamma} = \Lambda, \tilde{P} = I, \tilde{\Phi} = P'\Phi$, $\tilde{\mathbf{T}} = P'\mathbf{T}$ and if the theorem holds for $\tilde{\mathbf{U}}_n$ we obtain

$$n^{\beta/2}\tilde{\mathbf{U}}_n \sim \eta((\Lambda - (\beta_+/2)\,I)^{-1}P'\mathbf{T}, M)$$

which implies $n^{\beta/2}\mathbf{U}_n \sim \eta(P(\Lambda-(\beta_+/2)\,I)^{-1}\,P'\mathbf{T}, PMP') = \eta$. Hence we may assume $\Gamma = \Lambda$, $P = 1$. Next we observe that $n^{\beta/2}\mathbf{U}_n \sim (n-1)^{\beta/2}\,\mathbf{U}_n = \tilde{\mathbf{U}}_n$ (say) we have to prove $\tilde{\mathbf{U}}_n \sim$, but $\tilde{\mathbf{U}}_n$ satisfies (6) with $\tilde{\beta} = \beta_+ = 0$ and $\tilde{\Gamma}_n$ satisfying

$$I - n^{-\alpha}\tilde{\Gamma}_n = [n/(n-1)]^{\beta/2}\,(I-n^{-\alpha}\Gamma_n)$$

whence it is easy to verify that $\tilde{\Gamma}_n \to \Lambda-(\beta_+/2)\,I$ and the theorem holds if it holds for $\beta = 0$. Now for any $\epsilon > 0$, by Egorov's theorem, we can change $\Gamma_n, \Phi_n, \mathbf{T}_n$ on a set of probability at most ϵ and obtain $\tilde{\Gamma}_n, \tilde{\Phi}_n, \tilde{\mathbf{T}}_n$ converging uniformly on Ω to their limits Γ, Φ, $\mathbf{T}$ (the last only if $\mathbf{T}_n \to \mathbf{T}$). We can do that so that $\tilde{\Gamma}_n, \tilde{\Phi}_n$ remain F_n-measurable, and we then define $\tilde{\mathbf{U}}_n$ by (6) with $\tilde{\Gamma}_n, \tilde{\Phi}_n, \tilde{\mathbf{T}}_n$ substituted for $\Gamma_n, \Phi_n, \mathbf{T}_n$. If then the theorem holds if the convergence of Γ_n, Φ_n, $\mathbf{T}_n$ is uniform, we obtain $\tilde{\mathbf{U}}_n \sim \eta$ and, because ϵ was arbitrary and

$$P\{\mathbf{U}_n = \tilde{\mathbf{U}}_n \quad \text{all} \quad n\} \geqslant 1-\epsilon, \mathbf{U}_n \sim \eta.$$

So we can assume $\Gamma_n \to \Lambda$, $\Phi_n \to \Phi$ uniformly and $E\|\mathbf{T}_n - \mathbf{T}\| \to 0$. With $\Phi_n \to \Phi$ uniformly we can change $\mathbf{V}_n$ to $\Phi_n\mathbf{V}_n$, Φ_n to I without changing U_n, and without violating (2), (3), (4) and the F_{n+1}-measurability of $\mathbf{V}_n$. Now suppose $\tilde{\mathbf{U}}_n$ is defined by (6) except that in the right-hand side $\tilde{\mathbf{U}}_n$ is substituted for $\mathbf{U}_n$ and a $\mathbf{Z}_n$ is subtracted. If then $\boldsymbol{\Delta}_n = \mathbf{U}_n - \tilde{\mathbf{U}}_n$, we obtain

$$\boldsymbol{\Delta}_{n+1} = (I-n^{-\alpha}\Gamma_n)\,\boldsymbol{\Delta}_n + \mathbf{Z}_n$$

and
$$\|\boldsymbol{\Delta}_{n+1}\| \leqslant |1-n^{-\alpha}[\lambda+o_u(1)]|\,\|\boldsymbol{\Delta}_n\| + \|\mathbf{Z}_n\|. \tag{9}$$

If $\mathbf{Z}_n = 0$ and $\boldsymbol{\Delta}_1 = \mathbf{U}_1$ then Problem 5 of Appendix 3 implies $\|\boldsymbol{\Delta}_n\| \to 0$, $\tilde{\mathbf{U}}_n \sim \mathbf{U}_n$ and thus we may assume $\mathbf{U}_1 = 0$. Next we set $\boldsymbol{\Delta} = 0$, $\mathbf{Z}_n = n^{-\alpha}(\mathbf{T}_n - \mathbf{T})$, we can take expectations in (9) and because $E\|\mathbf{Z}_n\| = o(n^{-\alpha})$, we obtain $E\|\boldsymbol{\Delta}_n\| \to 0$, $\tilde{\mathbf{U}}_n \sim \mathbf{U}_n$ and we may assume $\mathbf{T}_n = \mathbf{T}$. Now setting $\mathbf{Z}_n = n^{-\alpha}\mathbf{T}$ we obtain from (9) that $\|\boldsymbol{\Delta}_n\| = O(1)$ and from the relation preceding (9), $\boldsymbol{\Delta}_{n+1} = (I-n^{-\alpha}\Lambda)\,\boldsymbol{\Delta}_n + n^{-\alpha}\mathbf{T} + n^{-\alpha}(\Lambda-\Gamma_n)\,\boldsymbol{\Delta}_n$. The last term is $o(n^{-\alpha})$ and a coordinatewise application of Problem 5, Appendix 3 gives $\boldsymbol{\Delta}_n \to \Gamma^{-1}\mathbf{T}$. Hence $\mathbf{U}_n \sim \eta(\Gamma^{-1}\,\mathbf{T}, M)$ if and only if $\tilde{\mathbf{U}}_n \sim \eta(0, M)$, $\tilde{\mathbf{U}}_n$ satisfies all conditions of the theorem with $\mathbf{T} = 0$. Hence we can assume (8) holds.

We have now $\mathbf{U}_{n+1} = (I - n^{-\alpha}\Gamma_n)\,\mathbf{U}_n + n^{-\alpha/2}\mathbf{V}_n$ and because of (2), $E_{F_n}(I - n^{-\alpha}\Gamma_n)\,\mathbf{U}_n\mathbf{V}_n = 0$. Hence

$$E\|\mathbf{U}_{n+1}\|^2 = (1 - n^{-\alpha}(\lambda + o(1)))^2 E\|\mathbf{U}_n\|^2 + n^{-\alpha/2}o(1)$$
$$= |1 - 2n^{-\alpha}\lambda + o(1)|\,E\|\mathbf{U}_n\|^2 + n^{-\alpha/2}O(1)$$

and Problem 5, Appendix 3 implies $E\|\mathbf{U}_n\|^2 = \mathbf{o}(1)$. Setting them $\tilde{\mathbf{U}}_{n+1} = (I - n^{-\alpha}\Lambda)\,\tilde{\mathbf{U}}_n + n^{-\alpha/2}\mathbf{V}_n$ we obtain (9) for $\mathbf{\Delta}_n = \mathbf{U}_n - \tilde{\mathbf{U}}_n$ with $E\|\mathbf{Z}_n\| = n^{-\alpha}E\|\Gamma_n - \Lambda\|\,\|\mathbf{U}_n\| = o(n^{-\alpha})$, and $E\|\mathbf{\Delta}_n\| \to 0$, $\mathbf{U}_n \sim \tilde{\mathbf{U}}_n$ and we may even assume $\Gamma_n = \Lambda$,

$$\mathbf{U}_{n+1} = (I - n^{-\alpha}\Lambda)\,\mathbf{U}_n + n^{-\alpha/2}\mathbf{V}_n. \tag{10}$$

From here and by (2) we obtain

$$E\mathbf{U}_{n+1}\mathbf{U}'_{n+1} = (I - n^{-\alpha}\Lambda)\,E\mathbf{U}_n\mathbf{U}'_n(I - n^{-\alpha}\Lambda) + n^{-\alpha}\Sigma + o(n^{-\alpha})$$

and a coordinate-wise application of Problem 5, Appendix 3 gives

$$(E\mathbf{U}_n\mathbf{U}_n)^{(ij)} \to \Sigma^{(ij)}/(\Lambda^{(ii)} + \Lambda^{(jj)}), \quad \text{i.e.} \quad E\mathbf{U}_n\mathbf{U}'_n \to M.$$

For the third step we need the following standard relation (e.g. Feller (1966), proof of Theorem 1, XV. 6): If $\delta > 0$ then $e^{iy} = 1 + iy - y^2/2 + R$ with $|R| \leqslant \delta y^2 + y^2\chi\{y^2 \geqslant \delta^2\}$. We can clearly choose a sequence $\delta_n \to 0$, $\delta_n > 0$ such that (3) or (4) holds with δ_j substituted for r. With $y = n^{-\alpha/2}\mathbf{t}'\mathbf{V}_n$, $\delta = \|\mathbf{t}\|\delta_n, \mathbf{t} \in R^k$ we obtain

$$E_{F_n}e^{iy} = 1 - \tfrac{1}{2}n^{-\alpha}\mathbf{t}'\Sigma\mathbf{t} + \tfrac{1}{2}n^{-\alpha}\mathbf{t}'(\Sigma - E_{F_n}\mathbf{V}_n\mathbf{V}'_n)\,\mathbf{t} + R$$

with $E|R| \leqslant n^{-\alpha}\|\mathbf{t}\|^3\,\delta_n\sigma^2_{n,o} + n^{-\alpha}\|\mathbf{t}\|^2\sigma^2_{n,\delta_n}$ because

$$\chi\{n^{-\alpha}(\mathbf{t}'\mathbf{V}_n)^2 \geqslant \|\mathbf{t}\|^2\delta_n^2\} \leqslant \chi\{\|\mathbf{V}_n\|^2 \geqq n^{\alpha}\delta_n^2\}.$$

Hence, limiting our considerations to a fixed sphere

$$\|\mathbf{t}\| \leqslant \tau, \quad \tau > 0,$$

and noting that (2) implies $\sigma^2_{n,o} = O(1)$, we obtain

$$E|E_{F_n}[\exp(in^{-\alpha/2}\mathbf{t}'\mathbf{V}_n)] - 1 + \tfrac{1}{2}n^{-\alpha}\mathbf{t}'\Sigma\mathbf{t}| \leqslant \|\mathbf{t}\|n^{-\alpha}h_n \tag{11}$$

with $h_n = o(1) + \sigma^2_{n,\delta_n}$, so that either $h_n \to 0$ or $\alpha = 1$ and

$$\frac{1}{n}\sum_{j=1}^{n} h_j \to 0.$$

Now let us denote $B_n = I - n^{-\alpha}\Lambda$, $\varphi_n(\mathbf{t}) = Ee^{i\mathbf{t}U_n}$, $\psi_1(\mathbf{t}) = 1$, $\psi_{n+1}(\mathbf{t}) = \psi_n(\mathbf{B}_n\mathbf{t})(1-\frac{1}{2}n^{-\alpha}\mathbf{t}'\Sigma\mathbf{t})$. Then a rearrangement of terms, (11) and F_n-measurability of U_n give

$$|\varphi_{n+1}(\mathbf{t}) - \psi_{n+1}(\mathbf{t})| = |E\{[e^{i\mathbf{t}'B_n\mathbf{U}_n} - \psi_n(\mathbf{B}_n\mathbf{t})](1-\tfrac{1}{2}n^{-\alpha}\mathbf{t}'\Sigma\mathbf{t})$$

$$+ e^{i\mathbf{t}'B_n\mathbf{U}_n}[e^{i\mathbf{t}n^{-\alpha/2}\mathbf{V}_n} - 1 + \tfrac{1}{2}n^{-\alpha}\mathbf{t}'\Sigma\mathbf{t}]\}|$$

$$\leqslant |1-\tfrac{1}{2}n^{-\alpha}\mathbf{t}'\Sigma\mathbf{t}|\,|\varphi_n(B_n\mathbf{t}) - \psi_n(B_n\mathbf{t})| + \|\mathbf{t}\|n^{-\alpha}h_n.$$

The relation $|\varphi_j(\mathbf{t}) - \psi_j(\mathbf{t})| \leqslant \mathbf{\Delta}_j\|\mathbf{t}\|$ holds with $\mathbf{\Delta}_1 = 0$ for $j = 1$. If it holds for $j = n$ then, because

$$|\varphi_n(B_n\mathbf{t}) - \psi_n(B_n\mathbf{t})| \leqslant \Delta_n\|B_n\mathbf{t}\| \leqslant \Delta_n\|B_n\|t,$$

it holds with a $\Delta_{n+1} \in R$ for $j = n+1$, too. For sufficiently large n, $\|B_n\| = 1-n^{-\alpha}\lambda$, $|1-\frac{1}{2}n^{-\alpha}\mathbf{t}'\Sigma\mathbf{t}| \leqslant 1$ and Δ_{n+1} can be chosen as $\mathbf{\Delta}_{n+1} = (1-n^{\alpha}\lambda)\mathbf{\Delta}_n + n^{-\alpha}h_n$. Problem 4, Appendix 3 then implies that $\mathbf{\Delta}_n \to 0$. Because τ was arbitrary this means

$$\phi_n(\mathbf{t}) - \psi_n(\mathbf{t}) \to 0 \quad \text{for every } t.$$

The whole proof will be completed when we show that ψ_n converges to the characteristic function ψ corresponding to η. Suppose $\mathbf{V}_1, \mathbf{V}_2, \ldots$ are independent and $\eta(0, \Sigma)$ distributed. Then $\mathbf{U}_n$ is normal for every n, $E\mathbf{U}_n = 0$, $E\mathbf{U}_n\mathbf{U}_n' \to M$ by what has already been proved and thus $\mathbf{U}_n \sim \eta, \varphi_n \to \psi, \psi_n \to \psi$.

4. Applications

Once one has a distribution of a random variable, one can construct many interesting statistics to draw inferences. This is particularly easy to do when a normal density function is available. Suppose for example that one is interested in finding an unbiased estimate of the 80th percentile of a normal distribution $N(\mu, \sigma^2)$ where μ and σ^2 are both unknown. In other words, one is interested in estimating $\mu + k\sigma$ where k is a suitable real number (k is known). Such a question can be answered very easily by constructing unbiased estimates $\hat{\mu}$ of μ and $\hat{\sigma}$ of σ. Then $\hat{\mu} + k\hat{\sigma}$ is an unbiased estimate of $\mu + k\sigma$; or one may be interested in construction of a confidence interval for a parameter which is discussed below.

Confidence intervals.

In special cases exact probability statements for the procedures discussed in the preceding sections can be made, even when only a finite number of steps have been carried out. Suppose $M(x) = \alpha(x-\theta)$ with $\alpha > 0$ and α is known, and that, for each x, $Z(x)$ is a $N(0,\sigma)^2$-random variable (r.v.). As in Theorem 2 we can have

$$X_{n+1}-\theta = -\alpha^{-1}n^{-1}\sum_{j=1}^{n} Z_j = -\alpha^{-1}\bar{Z}_n, \tag{1}$$

where n is any finite positive integer $a_n = \alpha/n$, the Z_j's are independent and identically distributed, each having the distribution of a $N(0,\sigma^2)$-random variable. Hence $X_{n+1}-\theta$ is a $N(0,\sigma/n\alpha^2)$-random variable. Let $\Phi(t)$ denote the distribution function of $N(0,1)$-r.v. and Z_δ is such that

$$\Phi(Z_\delta)-\Phi(-Z_\delta) = 1-\delta.$$

Let σ^2 be a known constant. Then, for $0 < \delta < 1$, the interval

$$\left[X_{n+1}-\frac{Z_\delta\sigma}{\alpha\sqrt{n}},\quad X_{n+1}+\frac{Z_\delta\sigma}{\alpha\sqrt{n}}\right] \tag{2}$$

contains θ with probability $1-\delta$.

If σ^2 is not known then one can use the following estimate of σ^2

$$\hat{\sigma}_n^2 = (n-1)^{-1}\sum_{j=1}^{n}[y_j-(x_j-x_{n+1})]^2. \tag{3}$$

Substituting $y_j = Z_j+M(x_j) = Z_j+\alpha(x_j-\theta)$ and (1) in (3) we get

$$\hat{\sigma}^2 = (n-1)^{-1}\sum_{j=1}^{n}[\bar{Z}_j-\bar{Z}]^2, \tag{4}$$

thus it follows that $[(n-1)\hat{\sigma}_n/\sigma^2]$ has a χ^2_{n-1} distribution, independent of X_{n+1}. One can construct confidence interval

$$[X_{n+1}-t_{n-1,\delta}\hat{\sigma}_n/\alpha\sqrt{n}, X_{n+1}+t_{n-1,\delta}\hat{\sigma}_n/\alpha\sqrt{n}] \tag{5}$$

for θ having the probability $1-\delta$, where $t_{n-1,\delta}$ is such that $\Phi(t_{n-1,\delta}) - \Phi(-t_{n-1,\delta}) = 1-\delta$, where $\Phi(t)$ is the distribution function of a t-random variable with $n-1$ degrees of freedom.

Wasan [80] discusses other applications of estimation techniques.

5. Problems

1. For the stochastic-approximation plan defined in § 4 (*b*) of Chapter 4 prove that

$$\sqrt{n}(T_{\lambda_n} - T_\lambda) \xrightarrow{\mathscr{L}} N(0, (1-p)p^{-1}T_\lambda^{-2}) \quad \text{as} \quad n \to \infty,$$

i.e. $\sqrt{n}(T_{\lambda_n} - T_\lambda)$ is asymptotically normally distributed with mean 0 and variance $[(1-p)p^{-1}T_\lambda^{-2}]$, where $N(0, (1-p)p^{-1}T_\lambda^{-2})$ denotes the normal probability density function with mean 0 and variance $(1-p)p^{-1}T_\lambda^{-2}$. (Venter and Gastwirth).

2. Assume conditions of the quantal response problem discussed in § (5) of Chapter 4. Then prove for the recursive relation

$$X_{n+1} = X_n - \frac{f(0)}{\tau n}[Y(X_n) - \tfrac{1}{2}]$$

that $$\sqrt{n}(X_{n+1} - m) \xrightarrow{\mathscr{L}} N\left(0, \frac{1}{4}\frac{\tau^2}{[f(0)]^2}\right) \quad \text{as} \quad n \to \infty.$$

(Venter)

3. Assume conditions (1) (2) and (3) of § 3 (*a*) of Chapter 5 and let $\{Y_i(X)\}$, $i = 1, 2, \ldots, k$. Suppose also the following conditions are satisfied.

(i) $M_i(X) = \alpha_i + b_i X \quad (i = 1, 2, \ldots, k)$.

(ii) There exist positive constants $\mathbf{V}$ and $\overline{V}$ such that

$$\mathbf{V} < \sigma_i^2(X) < \overline{V} \quad \text{for all} \quad X, i = 1, 2, \ldots, k.$$

(iii) $\sigma_i^2(.)$ is continuous for $i = 1, 2, \ldots, k$.

(iv) $H_i(y/.)$ is Borel-measurable for all real y, $i = 1, 2, \ldots, k$.

(v) For all natural numbers r,

$$\sup_x E\{|Y_i(X) - M_i(X)|^r\} < \infty \quad (i = 1, 2, \ldots, k).$$

(vi) $na_n \to c > \frac{1}{2}(\min(b_1, \ldots, b_k))$,

then prove that $n^{\frac{1}{2}}(\mathbf{X}_n - \theta_n)$ is asymptotically normal with mean 0 and variance-covariance matrix $C'\Sigma C$. Prove that, moreover, the moments of $n^{\frac{1}{2}}(\mathbf{X}_n - \theta_n)$ converge to the moments of the limit distribution where C satisfies the relation

$$\mathbf{U}_n = C\mathbf{X}_n,$$

(where C is a $k\times k$ orthogonal matrix) and these exist constants λ_i, $i = 1, 2, \ldots, k$ such that

$$E\{u_{i_{n+1}}|u_{i_n}\} = u_{i_n} - a_n\lambda_i u_{i_n} \quad \text{for } i = 1, 2, \ldots, k \quad \text{for all } n.$$

$$\Sigma = (c^2 V_{ij}(0)/[c(\Delta_i + \Delta_j) - 1]),$$

$$V_{ij}(0) = \begin{cases} 0 & \text{for} \quad i = 1 \quad \text{or} \quad j = 1, \\ \sum_{m=1}^{k} C'_{im} C'_{jm} \sigma_m^2(\theta_m) & \text{for} \quad i, j = 2, \ldots, k. \end{cases}$$

$$C'_{ij} = \left(\sum_{j=1}^{k} C_{ij}/k\right) - C_{ij}.$$

C_{ij} is an element of ith row and jth column of C. The elements of C can be developed as follows:

$$u_n = \sum_1^k C_i X_{i_n} = \sum_{i=1}^k C_i$$

there exists Δ such that

$$E[u_{n+1}|u_n] = u_n - (a_n/_k)\,\Delta u_n\, kb_i C_i - b_i = \Delta C_i \qquad \text{for} \quad i = 1, 2, \ldots, k.$$

Let $\mathbf{C} = [C_1, \ldots, C_k]$ and (a_{ij}) be the $k\times k$ matrix with elements $a_{ij} = -b_i + \delta_{ij} kb_i$, where δ_{ij} is Kronecker δ. Then

$$(a_{ij})\,\mathbf{C}' = \Delta\mathbf{C}', \quad (a_{ij}) = (d_{ij})\,(\alpha_{ij}),$$

where $\quad d_{ii} = b_i,$

$$d_{ij} = 0 \quad \text{if} \quad j \neq i,$$

$$(\alpha_{ij}) = \begin{pmatrix} k-1 & -1 & -1 & -1 \\ -1 & k-1 & -1 & -1 \\ \cdots & \cdots & \cdots & \cdots \\ -1 & -1 & -1 & k-1 \end{pmatrix}.$$

(a_{ij}) has the eigenvalue zero and $k-1$ non-zero eigenvalues counting multiplicities. Let $\Delta_1 = 0, \Delta_2, \ldots, \Delta_k$ be the remaining eigenvalues counting multiplicity. One can choose orthogonal vectors $\mathbf{C}_j$, $j = 1, \ldots, k$ such that

$$(a_{ij})\,\mathbf{C}'_j = \Delta_j \mathbf{C}'_j.$$

Let these be the row vectors of matrix C. Thus C is defined.

(Epling)

4. (*a*) Suppose X_n is a type A_1 process, θ is a real number and each of $\alpha_1, \sigma^2, \epsilon_0, c$ and r_0 is a positive number such that

(i) If $x \neq \theta$ then $(x-\theta)(M(x)-\alpha) > 0$.

(ii) M is differentiable at θ and $M'(\theta) = \alpha_1$.

(iii) $\sup_x \left[\dfrac{M(x)}{1+x}\right] < \infty$.

(iv) If $0 < \delta_1 < \delta_2 < \infty$ then $\inf |M(x)-\alpha| > 0$ for

$$\delta_1 \leqslant |x-\theta| < \delta_2.$$

(v) V is bounded and is continuous at θ and $v(\theta) = \sigma^2$.

(vi) If r is in N then $\sup E|Y(x)-M(x)|^r < \infty$ for $|x-\theta| < \epsilon_0$.

(vii) $H(y|\,.)$ is Borel measurable for each y in R.

(viii) $na_n \to \dfrac{1}{2\alpha_1} r_n \to r_0$;

then prove that $n^{\frac{1}{2}}(X_n-\theta)$ is asymptotically normal

$$[0, \sigma^2 c^2/r_0(2\alpha_1 c - 1)].$$

(Burkholder)

(*b*) State conditions under which $\sqrt{n}(M(x)-M(\theta))$ is asymptotically normal and prove the result.

5. For the illustrative example discussed in Chapter 3, §5, prove asymptotic normality of the stochastic approximation procedure.

6. Let h be a function defined on R. Suppose there is a $\theta \in R$ such that for any closed finite interval $I \subset (-\infty, 0)$, $J \subset (0, \infty)$ we have $\sup h(I) < 0$ and $\inf h(J) > 0$. The function h has a bounded second derivative in a neighbourhood of θ,

$$d = h'(\theta) > 0$$

and there are $A, B \in R$ such that

$$|h(x)| \leqslant A|x-\theta| + B. \tag{1}$$

Let X_n, Y_n, Z_n be random variables,

$$F_n = \sigma(X_1, X_1, \ldots, Y_{n-1}, Z_1, \ldots, Z_{n-1}),^*$$

$$M_n = E_{F_n} Y_n, \quad N_n = E_{F_n} Z_n, \quad V_n = Y_n - M_n,$$

let
$$X_{n+1} = X_n - d_n n^{-1} Y_n, \tag{2}$$

$$M_n = \tfrac{1}{2}[h(X_n + c_n) + h(X_n - c_n)],$$
$$N_n = (2c_n)^{-1}[h(X_n + c_n) - h(X_n - c_n)] \tag{3}$$

with
$$c_n = cn^{-\gamma}, \quad \tfrac{1}{4} < \gamma < \tfrac{1}{2}. \tag{4}$$

Let
$$A_n = \frac{1}{n-1} \sum_{j=1}^{n-1} Z_j,$$

and, with $\vee$, $\wedge$ denoting maximum and minimum, respectively,

$$d_n = (C_1(\log(n+1))^{-1} \vee A_n^{-1}) \wedge C_2 n^{\alpha} \tag{5}$$

with $0 < C_1 < C_2$, $0 < \alpha < \frac{1}{2}$.

For some $\sigma > 0$ and some $C > 0$ let

$$E_{F_n}(Z_n - N_n)^2 \leqslant Cc_n^{-2}, \quad E_{F_n} V_n^2 \leqslant C \tag{6}$$

and
$$E_{F_n} \chi\{V_n^2 \geqslant r_n\} V_n^2 = s_{n,r}^2(X_n)$$

with
$$s_{n,0}^2(X_n) \to \tfrac{1}{2}\sigma^2, \quad \frac{1}{n}\sum_{j=1}^{n} s_{n,r}^2(X_n) \to 0 \tag{7}$$

for every sequence $X_n \to \theta$ and every $r > 0$.

Then $X_n \to \theta$, $d_n \to d^{-1}$ and $n^{\frac{1}{2}}(X_n - \theta)$ is asymptotically normal $(0, \frac{1}{2}\sigma^2 d^{-2})$. (Fabian)

7. Let k be a positive integer, X_n, $Y_n \in \mathbf{R}^k$, Σ, A, $P \in R^{k \times k}$, A positive definite, P orthogonal, $P'AP = \Lambda$ diagonal,

$$\lambda = \min \Lambda^{(ii)}, \quad \text{let} \quad m, \theta \in R^k, \quad 0 < \beta < 2\lambda a,$$

$$X_{n+1} = X_n - a_n Y_n, \tag{1}$$

$$F_n = \sigma(X_1, Y_1, Y_2, \ldots, Y_{n-1}), \quad M_n = E_{F_n} Y_n, V_n = c_n(Y_n - M_n),$$
$$a > 0, c > 0, C > 0,$$

$$a_n = an^{-1}, \quad c_n = cn^{-\gamma}, \quad \gamma = \tfrac{1}{2}(1-\beta), \tag{2}$$

$$X_n \to \theta, \quad C > \|E_{F_n} V_n V_n' - \Sigma\| \to 0, \quad \frac{1}{n}\sum_{j=1}^{n} \sigma_{j,r}^2 \to 0, \tag{3}$$

* I.e. the smallest σ-field with respect to which the indicated variables are measurable.

for every $r > 0$, with $\sigma_{j,r}^2$ as in Theorem 4; and let for X_n in a neighbourhood of θ

$$\|M_n - A(X_n - \theta) - n^{-\beta/2}m\| \leqq o(1)\,[n^{-\beta/2} + \|X_n - \theta\|]. \tag{4}$$

Then the asymptotic distribution of $n^{\beta/2}(X_n - \theta)$ is normal with mean $-a(aA - (\beta/2)\,I)^{-1}\,m$ and covariance matrix PMP' with $M^{(ij)} = a^2c^{-2}[P'\Sigma P]^{(ij)}/(a\Lambda^{(ii)} + a\Lambda^{(jj)} - \beta)$. (Fabian)

CHAPTER 7

THE APPROXIMATION FOR CONTINUOUS RANDOM PROCESSES

1. Introduction

In order to obtain a continuous version of the stochastic-approximation technique, one can replace a difference recursive iteration relation in the discrete case by a stochastic differential equation. We shall first prove a lemma which is useful in the proof of the main results. The techniques discussed in Theorems 1 and 2 can be exploited on an analogue computer. Then an application to a control problem is discussed.

In §3 a continuous Kiefer–Wolfowitz procedure for a random process on the lines of the work of Dupač [27] is developed which has a practical utility which can be noticed from its application.

2. A continuous version of stochastic approximation

We shall follow the method of Driml and Nedoma [25] which has uses in analogue computation. First consider the following lemma.

LEMMA (Driml–Nedoma). *Let $g(t)$ be a continuous real valued function such that*

$$\lim_{t\to\infty} \frac{1}{t}\int_0^t g(\tau)\,d\tau = m > 0 \tag{1}$$

exists. Let ϵ be a given number such that $0 < \epsilon < m$ and let $T > 1$ be such that for all $t \geqslant T$ the inequality

$$\left|\frac{1}{t}\int_0^t g(\tau)\,d\tau - m\right| < \epsilon \tag{2}$$

holds.

Then, for $v \geqslant T$ we have

$$\int_u^v a(\tau)\, g(\tau)\, d\tau > C + (m-\epsilon)\log v, \tag{3}$$

where C is a constant which does not depend on v, $a(\tau) = 1$ for $\tau \leqslant 1$ and $a(\tau) = 1/\tau$ for $\tau > 1$.

If $v \geqslant u \geqslant T$, then

$$\int_u^v a(\tau)\, g(\tau)\, d\tau > -2\epsilon + (m-\epsilon)\log\frac{v}{u} \geqslant -2\epsilon. \tag{4}$$

Proof. If $u < 1$, then

$$\int_u^v a(\tau)\, g(\tau)\, d\tau = \int_u^1 g(\tau)\, d\tau + \int_1^v a(\tau)\, g(\tau)\, d\tau$$

and the first member on the right-hand side remains constant for every $v \geqslant T$. Thus, it is sufficient to prove (3) for $u \geqslant 1$ only. In this case we obtain by integration by parts

$$\begin{aligned}\int_u^v a(\tau)\, g(\tau)\, d\tau &= \int_u^v \frac{1}{\tau} g(\tau)\, d\tau \\ &= \frac{1}{\tau}\int_0^\tau g(s)\, ds\Big|_{\tau=u}^{v} + \int_u^v \frac{1}{\tau^2}\int_0^\tau g(s)\, ds\, d\tau \\ &= \frac{1}{v}\int_u^v g(s)\, ds - \frac{1}{u}\int_0^u g(s)\, ds + \int_u^v \frac{1}{\tau}\cdot\frac{1}{\tau}\int_0^\tau g(s)\, ds\, d\tau.\end{aligned}$$

If $u < T$, we have

$$\begin{aligned}\int_u^v \frac{1}{\tau}\cdot\frac{1}{\tau}\int_0^\tau g(s)\, ds\, d\tau &= \int_u^T \frac{1}{\tau^2}\int_0^\tau g(s)\, ds\, d\tau \\ &\quad + \int_T^v \frac{1}{\tau}\cdot\frac{1}{\tau}\int_0^\tau g(s)\, ds\, d\tau > D + \int_T^v \frac{1}{\tau}(m-\epsilon)\, d\tau \\ &= D + (m-\epsilon)\log\frac{v}{T}.\end{aligned}$$

Thus

$$\begin{aligned}\int_u^v a(\tau)\, g(\tau)\, d\tau &> m - \epsilon - \frac{1}{u}\int_0^u g(s)\, ds \\ &\quad + D - (m-\epsilon)\log T + (m-\epsilon)\log v \\ &= C + (m-\epsilon)\log v.\end{aligned}$$

If $u \geqslant T$, then

$$\int_u^v \frac{1}{\tau} \cdot \frac{1}{\tau} \int_0^\tau g(s)\, ds\, d\tau > \int_u^v \frac{1}{\tau}(m-\epsilon)\, d\tau = (m-\epsilon)\log\frac{v}{u},$$

so that

$$\int_0^v a(\tau)\, g(\tau)\, d\tau > m-\epsilon-m-\epsilon+(m-\epsilon)\log\frac{v}{u}$$

$$= -2\epsilon+(m-\epsilon)\log\frac{v}{u}$$

and (3) together with (4) is proved.

We shall now prove the first main result of this section, a theorem on the convergence of stochastic approximation, defined below.

THEOREM 1 (Driml–Nedoma). Let $(\Omega, \mathfrak{S})$ be a measurable space with a complete probability measure μ. Let $f(\omega, t, x)$ be a random function of two real parameters t and x,

$$0 \leqslant t < \infty, \quad -\infty < x < \infty,$$

satisfying the following conditions:

(i) $\mu\{\omega : f(\omega, t, x)$ is continuous with respect to t and x simultaneously$\} = 1$;

(ii) for every t

$\mu\{\omega$: $x_1 < x_2$ implies $f(\omega, t, x_1) \leqslant f(\omega, t, x_2)$, $x_1, x_2 \in (-\infty, \infty)\} = 1$;

(iii) there exists a function $r(x)$ (called a regression function) such that

$$\mu\left\{\omega: \lim_{t\to\infty} \frac{1}{t}\int_0^t f(\omega, \tau, x)\, d\tau = r(x) \quad \text{for every } x\right\} = 1.$$

Let us set $\theta = \{x : r(x) = 0\}$.

Then the solution $x(\omega, t)$ of the differential equation

$$\frac{dx\langle\omega, t\rangle}{dt} = -a(t) f(\omega, t, x\langle\omega, t\rangle), \tag{5}$$

where

$$a(t) = \frac{1}{t}$$

exists with probability one and if $\theta \neq \phi$ the null set, then

$$\mu\{\omega: \lim_{t\to\infty} \rho(x\langle\omega, t\rangle, \theta) = 0\} = 1, \tag{6}$$

where ρ denotes the ordinary metric on the real line.

Proof. Let us denote by Ω_1 the set $\{\omega: f(\omega, t, x)$ is continuous with respect to t and x simultaneously; for every t and every x_1 and x_2, $x_1 < x_2$ implies

$$f(\omega, t, x_1) \leqslant f(\omega, t, x_2); \tag{7}$$

$$\lim_{t\to\infty} \frac{1}{t}\int_0^t f(\omega, \tau, x)\, d\tau = r(x) \quad \text{for every} \quad x\}.$$

The assumptions of the theorem yield immediately $\mu(\Omega_1) = 1$. The continuity and the monotony of the stochastic process $f(\omega, t, x)$ with probability one imply that the regression function $r(x)$ is continuous and non-decreasing as well. Consequently, θ is a closed interval, degenerate or not. Let us set $\theta = [a, b]$ (a and/or b can be equal to $-\infty$ and/or ∞ respectively). The almost sure solvability of the differential equation (5) follows immediately from the assumption (i).

In the remainder of the proof, we shall consider one fixed element ω from the set Ω_1. From (7) it follows that the solution of the differential equation (5) exists and, of course, is a continuous function of t. First of all, we shall prove that for every t_1 and every $\epsilon > 0$ there exists a point $t_2 > t_1$ such that

$$\rho(x\langle\omega, t_2\rangle, \theta) < \epsilon. \tag{8}$$

If $\rho(x\langle\omega, t_1\rangle, \theta) < \epsilon$ then the inequality (8) follows from the continuity of the trajectory $x\langle\omega, t\rangle$. If $\rho(x\langle\omega, t_1\rangle, \theta) \geqslant \epsilon$, we shall set $x\langle\omega, t_1\rangle = c$. We shall investigate only the case $c > b$, the remaining case being completely analogous.

Let us suppose that the point t_2 satisfying (8) does not exist, i.e. let us suppose that

$$x\langle\omega, t\rangle \geqslant b+\epsilon \quad \text{for} \quad t > t_1. \tag{9}$$

Integrating (5) we obtain using (ii)

$$\begin{aligned} x\langle\omega, t\rangle - x\langle\omega, t_1\rangle &= -\int_{t_1}^t a(\tau) f(\omega, \tau, x\langle\omega, \tau\rangle)\, d\tau \\ &\leqslant -\int_{t_1}^t a(\tau) f(\omega, \tau, b+\epsilon)\, d\tau. \end{aligned}$$

Condition (iii) and the formula (3) furnish the existence of a number T such that for all $t \geqslant T$

$$x\langle\omega, t\rangle - x\langle\omega, t_1\rangle < -C - (r(b+\epsilon) - \delta)\log t,$$

where $0 < \delta < r(b+\epsilon)$. Thus $\lim_{t\to\infty} x(\omega, t) = -\infty$ which contradicts our assumption (9).

Now, we shall prove that $\lim_{t\to\infty} \rho(x\langle\omega, t\rangle, \theta) = 0$. We shall suppose that it is not so, i.e. that there exists a positive number η such that for every t we can find a number t' for which

$$\rho(x(\omega, t'), \theta) > \eta.$$

Following from this assumption, and from the continuity of the function $x\langle\omega, t\rangle$, it follows from the proceeding part of the proof that either:

(a) for each t there exists a point $t'' > t$ such that

$$x\langle\omega, t''\rangle = b+\eta,$$

or

(b) for each t there exists a point $t'' > t$ such that

$$x\langle\omega, t''\rangle = a-\eta.$$

We shall prove that a contradiction can be deduced from the assertion (a). The same holds for the assertion (b), but, the proof being analogous will be omitted.

Let us denote by Δ the positive number

$$\min\left(r(b+\tfrac{1}{2}\eta), \tfrac{1}{4}\eta\right). \tag{10}$$

Then we can choose such a number $t_3 > 1$ that for all $t \geqslant t_3$ we have

$$\left| r(b+\tfrac{1}{2}\eta) - \frac{1}{t}\int_1^t f(\omega, \tau, b+\tfrac{1}{2}\eta)\, d\tau \right| < \Delta.$$

By (a) and by (8) there exists a point $t_4 > t_3$ satisfying

$$x\langle\omega, t_4\rangle = b+\tfrac{1}{2}\eta. \tag{11}$$

Now, by (a) we can find a point t_6 such that $t_6 > t_4$ and that

$$x\langle\omega, t_6\rangle = b+\eta. \tag{12}$$

Finally we shall denote by t_5 the greatest number satisfying $t_5 < t_6$ and

$$x\langle\omega, t_5\rangle = b+\tfrac{1}{2}\eta. \tag{13}$$

The existence of such a number t_5 follows from the continuity of the trajectory $x\langle\omega, t\rangle$. For all t such that $t_5 < t \leqslant t_6$ we then have

$$x\langle\omega, t\rangle > b+\tfrac{1}{2}\eta.$$

Now, integrating both sides of the differential equation (5) from t_5 to t_6, we obtain

$$x\langle\omega, t_6\rangle - x\langle\omega, t_5\rangle = \tfrac{1}{2}\eta = -\int_{t_5}^{t_6} a(\tau) f(\omega, \tau, x\langle\omega, \tau\rangle)\, d\tau$$

$$\leqslant -\int_{t_5}^{t_6} a(\tau) f(\omega, \tau, b + \tfrac{1}{2}\eta)\, d\tau.$$

Using again the lemma and (10) we have

$$\tfrac{1}{2}\eta < 2\Delta - \{r(b+\tfrac{1}{2}\eta) - \Delta\} \log \frac{t_6}{t_5} \leqslant 2\Delta \leqslant \tfrac{1}{2}\eta$$

and this is a contradiction.

Thus, for the above chosen $\omega \in \Omega_1$ the relation

$$\lim_{t\to\infty} \rho(x\langle\omega, t\rangle, \theta) = 0 \tag{14}$$

is proved and because this ω was chosen arbitrarily, (14) holds for every $\omega \in \Omega_1$, so that (5) and consequently the whole Theorem 1 is proved. Gardner [41] discusses an application of this theorem to a control problem.

In the second main result of this section we shall investigate a process $f(\omega, t, x)$ of additive form.

THEOREM 2 (Driml–Nedoma). *Let* $(\Omega, \mathfrak{S})$ *be a measurable space with a complete probability measure* μ. *Let* $r(x)$ *be a continuous real function of a real variable such that there exists a number* ν *with the following properties*:

$$r(\nu) = 0,$$

$$r(x) < 0 \quad \text{for all} \quad x < \nu,$$

$$r(x) > 0 \quad \text{for all} \quad x > \nu,$$

$$\liminf_{x\to\infty} r(x) > 0, \quad \limsup_{x\to-\infty} r(x) < 0.$$

Let $h(\omega, t)$ be a stochastic process so that

$$\mu(\omega\colon h\langle\omega, t\rangle \text{ is continuous}) = 1$$

and
$$\mu\left\{\omega\colon \left[\lim_{t\to\infty} \frac{1}{t}\int_0^t h\langle\omega, \tau\rangle\, d\tau\right] = 0\right\} = 1.$$

Set $$f(\omega, t, x) = h\langle\omega, t\rangle + r(x),$$

then the solution $x\langle\omega, t\rangle$ of the differential equation

$$\frac{dx\langle\omega, t\rangle}{dt} = -a(t) f(\omega, t, x\langle\omega, t\rangle) \tag{5}$$

exists with probability one and we have

$$\mu\{\omega: \lim_{t\to\infty} x\langle\omega, t\rangle = \nu\} = 1.$$

Proof. As in the proof of Theorem 1 we shall denote by Ω_2 the set

$$\left\{\omega: h(\omega, t) \text{ is continuous: } \lim_{t\to\infty} \frac{1}{t}\int_0^t h\langle\omega, \tau\rangle\, d\tau = 0\right\}.$$

It is evident that $\mu(\Omega_2) = 1$.

The almost sure solvability of equation (5) follows from the continuity of $r(x)$ and from the almost sure continuity of $h\langle\omega, t\rangle$.

Let ω be an arbitrary element of Ω_2. First we shall prove that the trajectory $x\langle\omega, t\rangle$ converges to some number, finite or infinite, as $t \to \infty$. Let us suppose that this is not so. Then there exist two numbers $a < b$, both greater than or smaller than ν at the same time and such that for every t_0 there exist points $t'' > t' > t_0$ so that

$$x\langle\omega, y'\rangle = a, x\langle\omega, t''\rangle = b,$$

$$a < x\langle\omega, t\rangle < b \quad \text{for all } t \text{ satisfying} \quad t' < t < t''.$$

In further considerations we shall investigate only the case $a > \nu$. The case ν can be examined analogously.

As already said, we suppose $a > \nu$. Let M be equal to the number $\min\limits_{x\in[a,b]} r(x)$. Evidently $M > 0$. Further we shall write

$$\delta = \min\left(M, \frac{b-a}{2}\right). \tag{15}$$

Let $t_1 > 1$ be such a number that for all $t \geqslant t_1$

$$\left|\frac{1}{t}\int_0^t h\langle\omega, \tau\rangle\, d\tau\right| < \delta.$$

Let us choose numbers $t_3 > t_2 > t_1$ so that

$$x(\omega, t_2\rangle = a, \quad x\langle\omega, t_3\rangle = b,$$

$$a < x\langle\omega, t\rangle < b \quad \text{for all } t \text{ satisfying} \quad t_2 < t < t_3.$$

By the integration of both sides of the differential equation (5) from t_2 to t_3 we obtain

$$\begin{aligned} b-a &= x\langle\omega,t_3\rangle - x\langle\omega,t_2\rangle \\ &= -\int_{t_2}^{t_3} a(\tau)\,[h\langle\omega,\tau\rangle]\,d\tau + \int_{t_2}^{t_3} a(\tau) r\,[x\langle\omega,\tau\rangle]\,d\tau \\ &\leqslant -\int_{t_2}^{t_3} a(\tau)\,[h\langle\omega,\tau\rangle + M]\,d\tau. \end{aligned}$$

The stochastic process $h\langle\omega,t\rangle + M$ has its time mean value equal to M for the considered $\omega \in \Omega_2$ and is therefore positive. Thus we can apply formula (4) from which together with (15) we get $b-a < 2\delta \leqslant b-a$, and this is the desired contradiction. Thus the trajectory $x\langle\omega,t\rangle$ converges for $t \to \infty$.

It remains to prove that it converges to the zero-point ν. Let us suppose the contrary, namely

$$\lim_{t\to\infty} x\langle\omega,t\rangle < \nu \quad \text{or} \quad \lim_{t\to\infty} x\langle\omega,t\rangle > \nu.$$

We shall again investigate only the second case.

The assumption $\lim_{t\to\infty} x\langle\omega,t\rangle > \nu$ implies that for a given number $\eta > 0$ for which $\eta < \lim_{t\to\infty} x\langle\omega,t\rangle - \nu$ there exists a point t_4 such that for all $t > t_4$

$$x\langle\omega,t\rangle > \nu + \eta. \tag{16}$$

Let us denote by N the positive number $\inf_{x \geqslant \nu+\eta} r(x)$. Let us choose a number Δ so that $0 < \Delta < N$ and let t_5 be a point such that for all $t \geqslant t_5$

$$\left|\frac{1}{t}\int_0^t h\langle\omega,\tau\rangle\,d\tau\right| < \Delta.$$

Integrating both sides of the differential equation (5) from t_5 to t and using (16) and (4) we conclude that

$$\begin{aligned} x\langle\omega,t\rangle - x\langle\omega,t_5\rangle &= \int_{t_5}^{t} a(\tau)\,[h\langle\omega,\tau\rangle + r(x\langle\omega,\tau\rangle)]\,d\tau \\ &\leqslant -\int_{t_5}^{t} a(\tau)\,[h\langle\omega,\tau\rangle + N]\,d\tau \\ &< 2\Lambda - (N-\Delta)\log\frac{t}{t_5}. \end{aligned}$$

Thus we have $\lim_{t\to\infty} x\langle\omega, t\rangle = -\infty$, which contradicts the assumption $x\langle\omega, t\rangle > \nu+\eta$. Therefore $\lim_{t\to\infty} x\langle\omega, t\rangle = \nu$ and since the point ω was chosen arbitrarily from the set Ω_2, Theorem 2 is completely proved.

3. A Kiefer–Wolfowitz procedure for continuous random processes

We shall follow the method of Dupač [27] to construct this procedure, using the following notation.

We regard the k parameters $x_1, x_2, \ldots, x_k$ as the components of a k-dimensional vector $\mathbf{x}$. The basis for the space will be the unit vectors $\mathbf{e}_1, \mathbf{e}_2, \ldots, \mathbf{e}_k$, $\mathbf{e}_i$ denoting a unit value of x_i and zero values for the other $(k-1)$ parameters. We will denote the usual Euclidean norm and inner product by $\|\mathbf{x}\|$ and $\langle \mathbf{x}, \mathbf{y}\rangle$ respectivly. We denote the regression function by

$$M(\mathbf{x}) = E\{Y_t(\mathbf{x})\} \tag{1}$$

and denote by $\boldsymbol{\theta}$ the vector parameter value for which M is a minimum. Let

$$Y_{i,t}[\mathbf{x}, c(t)] = Y_t[\mathbf{x}+c(t)\,\mathbf{e}_i] - Y_t[\mathbf{x}-c(t)\,\mathbf{e}_i] \tag{2}$$

in which $c(t)$ is a positive function whose properties will be described later. The minimum seeking approximation procedure is then defined by

$$(d/dt)\,X_{i,t} = -a(t)\,I_{i,t}\,c^{-1}(t)\,Y_{i,t}[\mathbf{X}_t, c(t)] \tag{3}$$

and $$X_{i,0} = x_i(0) \quad (i = 1, 2, \ldots, k),$$

in which $x_i(0)$ is the initial value of the ith parameter and

$$I_{i,t} = 1 - G_i^+[X_{i,t}]\,F_i^+[Y_{i,t}] - G_i^-[X_{i,t}]\,F_i^-[Y_{i,t}], \tag{4}$$

$$G_i^+(x) = 0,\ x \leqslant b_i-\delta \qquad G_i^-(x) = 0,\ x \geqslant a_i+\delta \tag{5}$$

$= 1,\ x = b_i,$ $\qquad$ $= 1,\ x = a_i,$

= monotone and of bounded derivative on $[b_i-\delta, b_i]$ $\qquad$ = monotone and of bounded derivative on $[a_i, a_i+\delta]$

and $$F_i^+(y) = 1 - y/\epsilon_y c(t), \quad F_i^-(y) = 1 + y/\epsilon_y c(t) \tag{6}$$

in which ϵ_y is a positive constant: we will later place a suitable bound on ϵ_y. For convenience in the sequel we will denote by $\mathbf{Y}_t$ the vector whose ith component is $\mathbf{Y}_{i,t}$ and by $\mathbf{Z}_t$ the vector whose ith component is $I_{i,t} Y_{i,t}$. We also define

$$\mathbf{M}_{c(t)}(\mathbf{x}) = E\{\mathbf{Y}_t(\mathbf{x})\} c^{-1}(t) \tag{7}$$

and $$\mathbf{Q}_{c(t)}(\mathbf{x}) = E\{\mathbf{Z}_t(\mathbf{x})\} c^{-1}(t). \tag{8}$$

Note that these quantities are defined in terms of the parameter x and not the random variable X_t.

THEOREM 3 (Sakrison). We now make the following assumptions:

(i) $Y_t(\mathbf{x}) = \sum_{j=1}^{N} g_j(x) V_{j,t}$, $N < \infty$ in which the $V_{i,t}$ are ergodic random processes which are bounded in magnitude with probability one and the g_j are functions whose second partial derivatives with respect to the x_i are bounded for all $x \in A$.

(ii) $$\langle \operatorname{grad} M(\mathbf{z})|_{\mathbf{z}=\mathbf{x}}, \mathbf{x} - \boldsymbol{\theta} \rangle \geqslant K \|\mathbf{x} - \boldsymbol{\theta}\|^2,$$

$$\|\operatorname{grad} M(\mathbf{x})\|^2 \leqslant K_1 \|\mathbf{x} - \boldsymbol{\theta}\|^2 \quad \text{all } \mathbf{x} \in A,$$

$$0 < K_0 \leqslant K_1 < \infty,$$

$$|\partial^3 M / \partial x_i^3| \leqslant P \quad \text{all} \quad x \in A \quad (i = 1, 2, \ldots, k),$$

$$a_i + 2\delta \leqslant \theta_i \leqslant b_i - 2\delta \quad (i = 1, 2, \ldots, k).$$

(iii) Let $D_{t+\rho}$ be any one of the random processes

$$V_{e,t+\rho} V_{m,t+\rho} (e, m = 1, 2, \ldots, N) \quad \text{or} \quad V_{e,t+\rho} (e = 1, 2, \ldots, N)$$

and let F_t be any bounded functional on the processes

$$V_{e,\tau} (e = 1, 2, \ldots, N), \tau \leqslant t \quad \text{and} \quad R_{FD}(\rho) = E\{(F_t - E\{F_t\})(D_{t+\rho} - E\{D_{t+\rho}\})\}$$

then we require for all $\rho \geqslant \rho_0, \rho_0 < \infty$

(iv) $|R_{FD}(\rho)| \leqslant \sigma_F \sigma_D (K_2/\rho^2) \quad (K_2 < \infty)$,

$$\int_0^\infty a(t)\, d\tau = \infty, \quad \int_0^\infty a(t)\, c^2(t)\, dt < \infty,$$

$$\int_0^\infty a(t)\, a(\tfrac{1}{2}t)\, dt < \infty \quad \text{and} \quad \int_1^\infty a(t)\, t^{-1}\, dt < \infty,$$

where

$$a(t) = a/(t+1)^{\alpha}, \quad c(t) = c/(t+1)^{\gamma}, \quad \tfrac{1}{2} < \alpha \leqslant 1, \quad \gamma > \tfrac{1}{2}(1-\alpha),$$

a and c are positive constants. (9)

If $\alpha = 1$, then assume

$$a > 4K_0^{-1}. \tag{10}$$

Then $\lim_{t\to\infty} E\{\|\mathbf{X}_t - \boldsymbol{\theta}\|^2\} = 0$ for all $\mathbf{x}(0) \in A$.

Proof. From equation (3) we have

$$\left.\begin{aligned} \frac{d}{dt}\|\mathbf{X}_t - \boldsymbol{\theta}\|^2 &= 2(\mathbf{X}_t - \boldsymbol{\theta})\frac{d\mathbf{X}_t}{dt}, \\ \left(\frac{d}{dt}\right)\|\mathbf{X}_t - \boldsymbol{\theta}\|^2 &= -2a(t)\,\langle \mathbf{X}_t - \boldsymbol{\theta}, c^{-1}(t)\,\mathbf{Z}_t(\mathbf{X}_t)\rangle. \end{aligned}\right\} \tag{11}$$

The right-hand side of this equation is bounded in magnitude with probability one for all t by condition (i) and so is $\|\mathbf{X}_t - \boldsymbol{\theta}\|^2$; thus by a theorem of Kolmogorov [51]

$$E\{(d/dt)\,\|\mathbf{X}_t - \boldsymbol{\theta}\|^2\} = (d/dt)\,E\{\|\mathbf{X}_t - \boldsymbol{\theta}\|^2\}. \tag{12}$$

For brevity we denote

$$b(t) = E\{\|\mathbf{X}_t - \boldsymbol{\theta}\|\}^2. \tag{13}$$

Adding and subtracting a term from the right-hand side of equation (11) and taking expected values yields

$$\begin{aligned} (d/dt)\,b(t) = {} & 2a(t)\,E\{\langle \mathbf{X}_t - \boldsymbol{\theta}, -\mathbf{Q}_c(\mathbf{X}_t)\rangle\} \\ & + 2a(t)\,E\{\langle \mathbf{X}_t - \boldsymbol{\theta}, \mathbf{Q}_c(\mathbf{X}_t) - c^{-1}(t)\,\mathbf{Z}_t(\mathbf{X}_t)\rangle\}. \end{aligned} \tag{14}$$

We now develop suitable bounds for the two terms on the right-hand side of this equation. First consider $M_c(x)$. By means of a Taylor's series we can express the ith component of this vector as

$$M_{c,i}(x) = 2[(\partial M(z)/\partial x_i)|_{z=x} + \tfrac{1}{6}c^2 R] \quad (|R| \leqslant P). \tag{15}$$

Thus, using condition (ii)

$$\langle \mathbf{x} - \boldsymbol{\theta}, -\mathbf{M}_c(\mathbf{x})\rangle \leqslant -2K_0\|\mathbf{x} - \boldsymbol{\theta}\|^2 + k^{\frac{1}{2}}P\tfrac{1}{3}c^2\|\mathbf{x} - \boldsymbol{\theta}\|. \tag{16}$$

The inner product

$$\langle \mathbf{x} - \boldsymbol{\theta}, -\mathbf{Q}_c(\mathbf{x})\rangle = \langle \mathbf{x} - \boldsymbol{\theta}, \mathbf{M}_c(\mathbf{x})\rangle + \sum_{i=1}^{k} (S_i^+ + S_i^-) \tag{17}$$

in which the quantities S_i^+ and S_i^- are given by

$$S_i^{\pm} = -(x_i - \theta_i)\, G_i^{\pm}(x_i)\{M_{c,i}(x)\,[\pm M_{c,i}(x)\,\epsilon_y^{-1} - 1] \pm \epsilon_y^{-1} c^{-2}(t)\,\sigma_{y_i}^2\}. \tag{18}$$

Note that by assumption (ii) and the definition of $G_i^{\pm}(x_i)$ that $x_i - \theta_i > \delta$ when $G_i^+(x_i)$ is non-zero and $x_i - \theta < -\delta$ when $G_i^-(x_i)$ is non-zero. The $M_{c,i}^2\epsilon_y^{-1}$ and $\sigma_{y_i}^2\epsilon_y^{-1}$ terms thus always make a negative contribution to S_i^+ or S_i^-. This allows an upper bound on the sum $\sum_{i=1}^{k}(S^+ + S_i^-)$. If we weaken this upper bound to ignore the $\sigma_{y_i}^2$ terms and combine the resultant with inequality (16) and assumption (ii), we obtain

$$\langle \mathbf{X} - \boldsymbol{\theta}, -\mathbf{Q}_c(\mathbf{x})\rangle \leqslant -2K_0\left[1 - (\epsilon_y/2K_0\sigma^2)\sum_{i=1}^{k}(b_i - a_i)\right]\|\mathbf{X} - \boldsymbol{\theta}\|^2 + K^{\frac{1}{2}}P\tfrac{1}{3}c^2\|\mathbf{X} - \boldsymbol{\theta}\|. \tag{19}$$

For ϵ_y suitably small and with a suitable definition of K_0, $K_0 > 0$ and $K_4 = \frac{1}{3}k^{\frac{1}{2}}P$ the inequality (20) follows.

$$\langle x - \boldsymbol{\theta}, -\mathbf{Q}_c(\mathbf{x})\rangle \leqslant -2K_0\|\mathbf{x} - \theta\|^2 + c^2K_4\|\mathbf{x} - \boldsymbol{\theta}\|. \tag{20}$$

Since this holds for any $\mathbf{x} \in A$ it also holds for any $\mathbf{X}_t$ generated by the approximation procedure up to time t; thus, substituting $\mathbf{X}_t$ into (20) and taking expected values of both sides we see that we can bound the first term on the right-hand side of equation (14) by

$$B_1(t) = -4a(t)\,K_0 b(t) + 2a(t)\,c^2(t)\,K_4 E\{\|\mathbf{X}_t - \boldsymbol{\theta}\|\}. \tag{21}$$

We now consider the second term on the right-hand side of equation (14).

The ith term in this inner product is

$$T_i = 2a(t)\,E\{\langle X_{i,t} - \boldsymbol{\theta}\rangle\,[Q_{c,i}(\mathbf{X}_t) - c^{-1}(t)\,Z_{i,t}(\mathbf{X})]\} \tag{22}$$

which, by using equations (2) to (6) and condition (i), may be expressed in the form

$$T_i = 2a(t)\sum_{j=1}^{N}\sum_{k=1}^{N} E\{f_{j,k}\langle \mathbf{X}_t, c(t)\rangle\,[V_{j,t}V_{k,t} - E\{V_{j,t}V_{k,t}\}]\} + \sum_{j=1}^{N} E\{f_j\langle \mathbf{X}_t, c(t)\rangle\,[V_{j,t} - E\{V_{j,t}\}]\}. \tag{23}$$

By condition (i) all of the $f_{jk}(x)$ and $f_j(x)$ appearing in this expression are bounded and possess bounded first partial derivatives with respect to all of the x_i for all $x \in A$ and all $t \geqslant 0$. Consider the e, mth term in this expression and for brevity denote $V_{e,t} V_{m,t}$ by D_t. Then this term can be written

$$T_{ime} = -2a(t) \sum_{q=1}^{k} \int_0^t a(\tau) E\left\{\frac{\partial f_{em}}{\partial X_q}(X_\tau, c(\tau)) \frac{Z_{q,\tau}}{c(\tau)} (D_t - E\{D_t\})\right\} d\tau \tag{24}$$

in which the order of integration and expectation has been exchanged since the process in the integrand is bounded in magnitude with probability one [51]. Now, since the processes inside the expectation are bounded with probability one for all e, m, and i, we can use condition (iii) to bound the expectation in the integrand in magnitude by

$$\left.\begin{aligned} L &= K_5 \qquad (t-1 \leqslant \tau \leqslant t), \\ &= K_{5/(t-\tau)^2} \quad (0 \leqslant \tau \leqslant t-1). \end{aligned}\right\} \tag{25}$$

Substituting this bound into equation (24), splitting the integral up into integrande over $[0, \frac{1}{2}t]$, $[\frac{1}{2}t, t-1]$, and $[t-1, t]$ and overbounding each integral, yields

$$|T_{ime}| \leqslant 2a(t)\,[K_6 a(\tfrac{1}{2}t) + K_7 \mu(t-1)\, t^{-1}]/KN(N+1) \qquad (K_6, K_7 < \infty). \tag{26}$$

in which
$$\mu(t) = 0 \quad t < 0,$$
$$= 1 \quad t \geqslant 0.$$

Thus the second term on the right-hand side of equation (14) may be bounded in magnitude by

$$B_2(t) = 2a(t)\,[K_6 a(\tfrac{1}{2}t) + K_7 \mu(t-1)\, t^{-1}] \tag{27}$$

combining this bound with that of equation (21) and inserting in equation (14) yields

$$\frac{db(t)}{dt} \leqslant -4a(t)\,K_0 b(t) + 2a(t)\,c^2(t)\,K_4 E\{\|\mathbf{X}_t - \boldsymbol{\theta}\|\} + 2a(t)\,[K_6 a(\tfrac{1}{2}t) + K_7 \mu(t-1)\, t^{-1}]. \tag{28}$$

Now, for any $\epsilon_t > 0$

$$E\{\|\mathbf{X}_t - \boldsymbol{\theta}\|\} \leqslant \epsilon_t + \epsilon_t^{-1} E\{\|\mathbf{X} - \boldsymbol{\theta}\|^2\} = \epsilon_t + b(t)\,\epsilon_t^{-1}. \tag{29}$$

Applying inequality (29) to inequality (28) for the choice

$$\epsilon_t = 2K_4 c^2(r)(K_0\epsilon)^{-1} \quad (0 < \epsilon < 4),$$

yields

$$(d/dt)\, b(t) + p(t)\, b(t) \leqslant q(t) \tag{30}$$

in which

$$p(t) = (4-\epsilon) K_0 a(t) > 0$$

and

$$q(t) = 2a(t)\,[K_6 a(t/2) + K_7 \mu(t-1)\, t^{-1}] + 4a(t)\, c^4(t)\, K_4 2(\epsilon K_0)^{-1} \geqslant 0.$$

Integrating both sides of equation (30) from 0 to t yields

$$b(t) + \int_0^t p(\tau)\, b(\tau)\, d\tau \leqslant \int_0^t q(\tau)\, d\tau + b(0), \quad b(0) = \|\mathbf{x}(0) - \theta\|^2. \tag{31}$$

Now, consider the integral equation

$$b_0(t) + \int_0^t p(\tau)\, b_0(\tau)\, d\tau = \int_0^t q(\tau)\, d\tau + b(0) \tag{32}$$

with solution

$$b_0(t) = b_0 \exp\left[-\int_0^t p(\tau)\, d\tau\right] + \int_0^\infty f_{[0,t]}(\tau)\, d\tau \tag{33}$$

in which

$$\left.\begin{aligned} f_{[0,t]}(\tau) &= \exp\left[-\int_\tau^t p(\xi)\, d\xi\right] q(\tau) \quad && (0 \leqslant \tau \leqslant t_1),\\ &= 0 && \text{elsewhere.}\end{aligned}\right\} \tag{34}$$

Now the non-negativeness of $p(t)$ and $g(t)$ and the continuity of $b(t)$ and $b_0(t)$ guarantee that, for any function $b(t)$ which satisfies inequality (31)

$$b(t) \leqslant b_0(t) \quad \text{for all} \quad t > 0 \tag{35}$$

(this is easily shown by assuming the contrary and reaching a contradiction). Thus we focus our attention on bounding $b_0(t)$. Now $0 \leqslant f_{[0,t]}(\tau) \leqslant q(\tau)$ for all t and τ greater than 0 and by condition (iv) $q(\tau)$ is integrable, thus by the general convergence theorem of Lebesgue and condition (iv)

$$\lim_{t\to\infty} b_0(t) = \lim_{t\to\infty} b(0) \exp\left[-\int_0^t p(\tau)\, d\tau\right] + \int_0^\infty \lim_{t\to\infty} f_{[0,t]}(\tau)\, d\tau = 0 \tag{36}$$

which completes the proof of Theorem 3.

4. An application to filter problem

The form of filter (or predictor) to be considered is shown in figure 1. The process V_t is the one observed and the process S_t is the one we desire to estimate. This form is general in that any

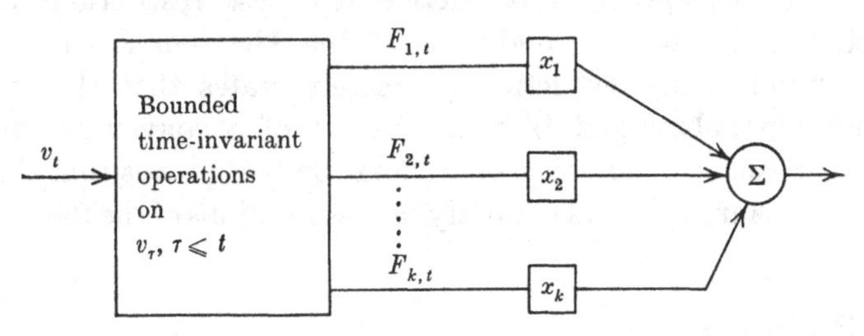

Filter output $Q_{t,(\mathbf{x})} = \sum_{j=1}^{k} x_j F_{j,t}$ = estimate of S_t

Fig. 1. Form of filter to be designed.

filter which operates on only a finite interval of the input process can be approximated arbitrarily closely by such a form (Cameron and Martin [11]). The parameters $x_1, x_2, \ldots, x_k$ are to be adjusted by the method described in order to minimize

$$M(\mathbf{x}) = E\{W[S_t - Q_t(\mathbf{x})]\}$$

in which W is some appropriate weighting function on the error, $S-Q$. (For a discussion of how this procedure can be mechanized by analogue simulation and of the restrictions involved, the reader is referred to Sakrison [65].)

The following restrictions on the processes involved and the error weighting function are sufficient to guarantee that conditions (i) to (iii) of the Theorem 3 are satisfied:

(*a*) The processes $F_{i,t} (i = 1, 2, \ldots, k)$ and S_t are jointly ergodic and bounded in magnitude with probability one.

(*b*) The correlation coefficient between any one of the F_i and any linear combination of the remaining F_j is unequal to ± 1.

(*c*) The function $W[e]$ is assumed to be a polynomial of degree $N, N < \infty$, and required to be 'strictly' convex in the sense that

$$W[\alpha a + (1-\alpha) b] \leqslant a W[a] + (1-\alpha) W[b] - E\alpha^2 |a-b|^2,$$
$$0 \leqslant \alpha \leqslant 1, \quad E \geqslant \epsilon > 0 \quad \text{for} \quad 0 \leqslant \alpha \leqslant \epsilon_0 > 0.$$

(*d*) Condition (iii) is assumed to hold for any random variable D_t of the form

$$D_t = (S_t)^{q_0} \prod_{j=1}^{k} (F_{j,t})^{q_j}, \sum_{j=0}^{k} q_j \leqslant 2N.$$

With the exception of restriction (*c*) these restrictions and their relation to the conditions of the Theorem 3 are quite straightforward. Restriction (*c*) merely states that the error weighting polynomial W must consist of a convex function plus a positive quadratic term. Sackrison [65] proves that these conditions are sufficient to satisfy the assumptions of the theorem.

5. Problems

1. Assume the conditions of Theorem 3 and let $\alpha = 1, \gamma \geqslant \frac{1}{4}$, then prove that

$$E\{\|\mathbf{X}_t - \mathbf{\theta}\|^2\} \leqslant \frac{K}{t+1}, \quad \text{all} \quad \mathbf{X}(0) \in A, K < \infty$$

and also that no choice of α and γ will yield a faster rate of convergence for all situations satisfying conditions (i) to (iv) of the Theorem 3.

2. Consider the following design of a filter (or predictor or model) of the form shown in the following figure

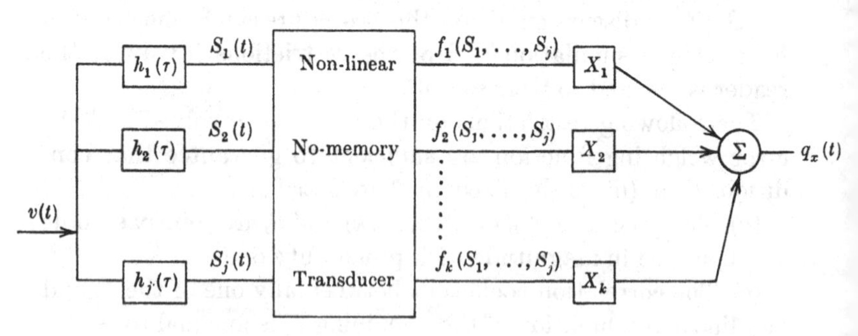

$$S_j(t) = \sum_{\tau=0}^{\infty} h_j(\tau) V(t-\tau) \quad \text{(Discrete time parameter case)}$$

$$S_j(t) = \int_0^{\infty} h(\tau) V(t-\tau)\, d\tau \quad \text{(Continuous time parameter case)}$$

Fig. 2

Let us assume that

(*a*) $V(t)$ and $d(t)$ are the outputs of stationary ergodic sources and are uniformly bounded in absolute magnitude for all t with probability one.

(*b*) $\sum_{t=0}^{\infty} |h_j| < \infty$, $i = 1, 2, \ldots, k$. The $f_i(t) = f_i[s_1(t), \ldots, s_j(t)]$ are continuous in $s_1, s_2, \ldots, s_j$ for $i = 1, 2, \ldots, k$.

(*c*) $P\left\{\left|\sum_{i=1}^{k} (X_i - \theta_i) f_{i,t}\right| \geqslant D\|\mathbf{X} - \theta\|\right\} \geqslant \epsilon$ for $\mathbf{X} \in X$, D and $\epsilon > 0$,

or in terms of one sample function, there exists an N_0 with the property that for $N > N_0$, $(n/2N+1) \geqslant \epsilon > 0$, where n is the number of occurrences

$$-N \leqslant t \leqslant N, \quad \text{of} \quad \left|\sum_{i=1}^{K} (X_i - \theta_i) f_i(t)\right| \geqslant D\|\mathbf{X} - \boldsymbol{\theta}\|, \mathbf{X} \in X.$$

(*d*) $W(e)$ is a polynomial in e of finite degree. ($W(e)$ possesses bounded continuous derivatives.)

(*e*) $W(e)$ is 'strictly convex', that is, there exists an $E > 0$ with the property that

$$W[\alpha a + (1-\alpha) b] \leqslant \alpha W(a) + (1-\alpha) W(b) - E\alpha|a-b|^2 \quad \text{for} \quad 0 \leqslant \alpha \leqslant \tfrac{1}{2}.$$

(*f*) The parameter adjustments are confined to a bounded, closed and convex set.

(*g*) Consider the random processes, $d(t)$ and $f_i(t)$, $i = 1, \ldots, k$; t may take on either discrete or continuous values. Let

$$F_1[f_1(\tau), \ldots, f_k(\tau), d\tau]$$

be any bounded continuous functional of $d(\tau)$ and the $f_i(\tau)$, $\tau \leqslant t$. Let $F_2[d(\tau), f_1(\tau), \ldots, f_k(\tau)]$ be any continuous functional on $d(\tau)$ and the $f_i(\tau)$, $t + \alpha \leqslant \tau \leqslant t + \alpha + T$, T fixed. Then let us make the following requirement on the correlation between F_1 and F_2 for large α

$$|R_{F_1F_2}| = |E\{(F_1 - \bar{F}_2)(F_2 - \bar{F}_2)\}| \leqslant \frac{k}{\alpha^2} \sigma_{F_1} \sigma_{F_2}$$

in which K is independent of F_1 and F_2. If $d(t)$ and the $f_i(t)$ are discrete time-parameter processes, then F_1 and F_2 are ordinary functions instead of functionals.

Then prove that this model satisfies the following conditions of a stochastic-approximation procedure for locating minimum of $M(\mathbf{X})$ $E\{W[d_t - q_{\mathbf{X},t}]\} = E[Y(\mathbf{X})]$.

(i) $E\{\|\mathbf{Y}_n\|^2 | \mathbf{X}_n(\mathbf{X}_1)\} \leqslant \|E\{\mathbf{Y}_n | \mathbf{X}_n(\mathbf{X}_1)\}\|^2 + S, S < \infty$;

(ii) (*a*) $K_0\|\mathbf{X} - \theta\|^2 \leqslant [(\text{grad}\, M)(\mathbf{X}) + (\mathbf{X} - \boldsymbol{\theta})]$,

(*b*) $\|(\text{grad}\, M)(\mathbf{X})\| \leqslant K_1\|X - \theta\|, k_1 > k_0 > 0$;

(iii) (*a*) $F_n(\mathbf{X}_1) = |E\{[\mathbf{X}_n - \boldsymbol{\theta}, \mathbf{Y}_n(\mathbf{X}_n) - C_n \mathbf{M}_{c_n}(\mathbf{X}_1)]\}| \leqslant S_1 \dfrac{a_{n/2}}{c_{n/2}}$,

(*b*) $|E\{\|E\{\mathbf{Y}_n | \mathbf{X}_n(\mathbf{X}_1)\}\|^2 - C_n^2 \|\mathbf{M}_{c_n}(\mathbf{X}_n)\|^2\} \leqslant S_2$ in which $S_1 < \infty$, $S_2 < \infty$ and S_i and S_2 are independent of $\mathbf{X}_1$ for all $\mathbf{X}_1 \in X$;

(iv) $\{a_n\}$ and $\{C_n\}$ are sequences of positive numbers satisfying

$$\sum_{n=1}^{\infty} a_n = \infty, \quad \sum_1^{\infty} (a_n)^2 < \infty, \quad \sum_1^{\infty} a_n C_n < \infty, \quad \sum_1^{\infty} (a_{n/C_n})^2 < \infty$$

$$\text{and} \quad \sum_1^{\infty} a_{n/C_n} \frac{a_{n/2}}{C_{n/2}} < \infty,$$

in which $a_{n/2}$ and $C_{n/2}$ are suitably interpolated for n odd;

(v) $\mathbf{X}$ is constrained to a bounded, closed convex set X, but is free to be varied inside X. (Sakrison)

CHAPTER 8

UP-AND-DOWN METHOD

1. Introduction

The up-and-down method is of use in fields where it is desirable to estimate a critical measurement for some response. In this method one performs an experiment at level y_0 (which might be an amount of a chemical or the height from which a bomb is dropped) and tests for a response (perhaps a chemical reaction or an explosion). If there is response one proceeds down a fixed amount d to $y_0 - d$ (d is a fixed number). If there is no response one goes up to $y_0 + d$. Thus the levels are

$$Y_n = \begin{cases} y_{n-1} - d & \text{if there was no response at } y_{n-1}, \\ y_{n+1} + d & \text{if there was a response at } y_{n-1}. \end{cases}$$

In the next section we formulate the problem mathematically to develop statistical inferential techniques.

2. Up-and-down method

The assumptions underlying the Dixon and Mood [23] method are as follows:

1. The natural variate should be transformed to a new variable which is distributed normally. In most fields of research it is well known, that with enough experience of the data we are dealing with, one can specify the nature of the distribution function of the random variable under question. Often it can be assumed in dosage-mortality experiments of this type, that the logarithm of the height (dosage-concentration) is normally distributed.

2. We should have some idea of the standard deviation of the normally distributed transformed variate. When the levels or heights used in the experiment are equally spaced, with a common distance d between them, Dixon and Mood [23] suggest

that d should be chosen approximately equal to the standard deviation. For the statistical analysis however Dixon and Mood show that even if the interval used is less than twice the standard deviation, simple analysis is possible.

3. Since large sample theory is used, the size of the sample must be 'large', for the analysis to be applicable at all. Dixon and Mood suggest that this analysis be restricted to sample sizes of over fifty observations.

Mathematical formulation.

Under the above assumptions let h be the height in an explosive experiment, so that $y = \log h$ is $N(\mu, \sigma^2)$ and d, a rough estimate of σ is the distance between testing heights. The experiment is performed as described, the first specimen being tested at the level which is closest to the anticipated mean. By the very nature of experimentation, the number of non-detonations at any given level can differ from the number of detonations at the next higher level by at most one. Let m_i non-detonations and n_i detonations have occurred at the ith level.

$$y_i = y_0 \pm id \quad (i = 0, 1, 2, \ldots), \tag{1}$$

y_0 being the initial level corresponding to height h_0. Let

$$N = \Sigma n_i \quad \text{and} \quad M = \Sigma m_i.$$

The probability of obtaining such a sample is

$$P(n, m/y_0) = k \prod_{i=-\infty}^{\infty} p_i^{n_i} q_i^{m_i}, \tag{2}$$

where p_i = probability of detonation at the ith level is

$$p_i = \frac{1}{\sigma\sqrt{2\pi}} \int_{-\infty}^{y_i} e^{-\frac{1}{2}\frac{(t-\mu)^2}{\sigma^2}} dt = 1 - q_i \tag{3}$$

and k is independent of μ and σ^2. Now

$$|n_i - m_{i-1}| = 0 \quad \text{or} \quad 1.$$

Hence either of the sets (n_i) or (m_i) summarizes practically all the information given by the sample. The smaller of N or M is chosen for the analysis ($N < M$ say). Now $M - N$ is expected to be small.

However, in the case where the initial level was rather poorly chosen, a certain number of observations have to be expended to get to the region of the mean and these observations contribute little to a precise location of the mean. This portion of the information is thus neglected and for simplicity of analysis, the likelihood function to be maximized is taken as

$$P(n, m/y_0, M-N) = k \prod_i (p_i q_{i-1})^{n_i}. \tag{4}$$

Applying the principle of maximum likelihood for the estimation of μ and σ^2, the derivatives of log P are equated to zero:

$$\frac{\partial \log P}{\partial \mu} = \Sigma n_i \left(\frac{z_{i-1}}{q_{i-1}} - \frac{z_i}{p_i}\right) = 0, \tag{5}$$

$$\frac{\partial \log P}{\partial \sigma} = \Sigma n_i \left(\frac{x_{i-1} z_{i-1}}{q_{i-1}} - \frac{x_i z_i}{p_i}\right) = 0. \tag{6}$$

where x_i is the standardized variable $(y_i - \mu)/\sigma$ and

$$z_i = \frac{1}{\sigma\sqrt{2\pi}} e^{-\frac{1}{2}\left(\frac{y_i-\mu}{\sigma}\right)},$$

the ordinate of the $N(\mu, \sigma^2)$ variate at y_i.

Set $w_0 = 1$

$$w_i = \begin{cases} \prod_{j=0}^{i-1} \frac{q_j}{p_j} & (i > 0), \\ \prod_{j=-i}^{1} \frac{p_j}{q_j} & (i < 0). \end{cases} \tag{7}$$

Now

$$\frac{E(n_i)}{p_i} = E(n_i + m_i)$$

$$= \text{expected number of observations on the } i\text{th level}$$

$$= \frac{E(n_{i+1})}{q_i} \quad \text{(approx.)}$$

$$\therefore E(n_i) = \frac{N w_i}{\sum_{-\infty}^{\infty} w_i}. \tag{8}$$

Close approximations for the roots of the maximum likelihood equations can be obtained when $d < 2\sigma$.

Consider

$$\alpha(u) = \frac{z(x)}{q(x)} - \frac{z\left(x+\frac{d}{\sigma}\right)}{p\left(x+\frac{d}{\sigma}\right)},$$

where

$$u = x+\frac{d}{2\sigma}.$$

This expression is nearly linear in u and similarly,

$$\beta(u) = \frac{xz(x)}{q(x)} = \frac{\left(x+\frac{d}{\sigma}\right)z\left(x+\frac{d}{\sigma}\right)}{p\left(x+\frac{d}{\sigma}\right)}$$

is nearly quadratic in u.

Thus setting

$$\mu_1 = \frac{1}{N}\Sigma n_i y_i, \tag{9}$$

$$\mu_2 = \frac{1}{N}\Sigma n_i y_i^2, \tag{10}$$

we have

$$E(\mu_1) = \mu - \frac{d}{2}, \tag{11}$$

$$E(\mu_2) = E^2(\mu_1) + \frac{d^2}{4} = \frac{\sigma^2 \Sigma w_i x_i^2}{\Sigma w_i}. \tag{12}$$

Thus Dixon and Mood show that the maximum likelihood estimates of μ and σ are

$$\hat{\mu} = y' + d\left(\frac{\Sigma i n_i}{N} \pm \tfrac{1}{2}\right), \tag{13}$$

y' being the normalized height corresponding to the lowest level on which the less frequent event (between detonations and non-detonations) occurs, the plus sign being used when the analysis is based on non-detonations and vice versa;

$$\hat{\sigma} = \text{the sample estimate of } \sigma$$

$$= 1{\cdot}620d\left[\frac{N[i^2 n_i - (\Sigma i n_i)^2]}{N^2} + 0{\cdot}029\right]. \tag{14}$$

The second derivatives of log P provide variances and covariances of the estimates $\hat{\mu}$ and $\hat{\sigma}$.

The following application has been investigated by Narayana [59].

3. Application to 'Rankits'

Let us apply the up-and-down method to the case of Rankits, where tolerance is assumed to be distributed rectangularly rather than normally. It is assumed that there are exactly $(n+1)$ levels, 0, 1, ..., n, the probability of explosion at the kth level being k/n, $k = 0, 1, \dots, n$. The probability of the next experiment being done at the $(k-1)$th level is k/n, and the probability of its being performed at the level $(k+1)$ is $1-(k/n)$. The experiment is thus seen to be a Markov chain with the stochastic matrix of order $n+1$.

$$\begin{pmatrix} 0 & 1 & 0 & 0 & 0 \\ 1/n & 0 & 1-(1/n) & 0 & 0 \\ 0 & 2/n & 0 & 1-(2/n) & 0 \\ \cdots & \cdots & \cdots & \cdots & \cdots \\ 0 & 0 & 0 & 0 & 1/n \\ 0 & 0 & 0 & 1 & 0 \end{pmatrix}.$$

This stochastic matrix characterizes the 'Ehrenfest Model of Diffusion' which represents diffusion with a central force. cf. Feller [36].

If $\{V_k\}$ is the stationary distribution, assuming we had started at the level k or any level differing from it by a multiple of 2 (e.g. $k-2$, $k-4$, etc.), let us find, that after $2m$ trials, where m is large, the probabilities V_{k-1}, V_{k+1}, V_{k+3} are all zero and after $(2m+1)$ trials, m being large, the probabilities V_{k-2}, V_k, V_{k+2} are all zero. The limiting distribution, after a long number of trials, is as shown below, depending on the level we start with and on whether an even or odd number of trials have elapsed.

$$\begin{array}{ll} V_{k+2} & V_{k+2} = 0; \\ V_{k+1} = 0, & V_{k+1}; \\ V_k, & V_k = 0; \\ V_{k-1} = 0, & V_{k-1}; \\ V_{k-2}, & V_{k-2} = 0; \\ V_{k-3} = 0, & V_{k-3}. \end{array}$$

Further,

$$V_k = \left(1-\frac{k-1}{n}\right)V_{k-1}+\frac{k+1}{n}V_{k+1} \quad (k = 1,\ldots,n-1),$$

$$V_0 = \frac{V_1}{n}, \quad V_n = \frac{V_{n-1}}{n},$$

$$\sum_{k=1}^{n} V_k = 1.$$

So that, after $2m$ trials,

$$V_{k-2}+V_k+V_{k+2}+\ldots = 1, \quad V_{k-1} = V_{k+1} = \ldots = 0$$

and after $(2m+1)$ trials

$$\ldots+V_{k-1}+V_{k+1}+\ldots = 1, \; V_{k-2} = V_k = V_{k+2} = \ldots = 0.$$

The required solution is, therefore,

$$V_k = \frac{\binom{n}{k}}{2^{n-1}}.$$

Let U_m denote the mth observation in an experiment using the up-and-down method in the case of Rankits. One can then obtain the values of $E(U_m)$, Var (U_m) and Cov $(U_m U_{m+1})$ for any m.

Evaluation of $E(U_m)$.

It is asumed that we have initially started at level j, the first observation being denoted by U_1. The second observation U_2 will be taken at level $(j-1)$ with probability j/n or at level $(j+1)$ probability $1-(j/n)$. Let U_m denote the mth observation taken at the level k (say) in such an experiment. Then

U_{m+1} is taken either at level $(k+1)$ with probability $1-(k/n)$ or at level $(k-1)$ with probability k/n.

Let us denote $E(U_m)$ by E_m

$$E(U_{m+1}) = (k+1)\left[1-\frac{k}{n}\right]+(k-1)\frac{k}{n} = 1+k\left(1-\frac{2}{n}\right),$$

i.e.

$$E_{m+1} = 1+E_m\left(1-\frac{2}{n}\right). \tag{1}$$

The solution of this difference equation is obtained by considering the corresponding homogeneous equation

$$E_{m+1} = E_m\left(1-\frac{2}{n}\right),$$

which yields for its solution

$$E_m = k\left(1-\frac{2}{n}\right)^{m-1}. \tag{2}$$

The value $E_m = c$ is a solution of (1) if $c = 1+c(1-2/n)$, i.e. if $c = n/2$.

Thus, we obtain

$$E_m = k\left(1-\frac{2}{n}\right)^{m-1}+\frac{n}{2} \tag{3}$$

as the solution of (1), where the constant k is determined from the initial condition $E_1 = j$.

$$\therefore\ k = \left(j-\frac{n}{2}\right)$$

and

$$E_m = \left(j-\frac{n}{2}\right)\left(1-\frac{2}{n}\right)^{m-1}+\frac{n}{2}. \tag{4}$$

Evaluation of variance of U_m.

Let us denote the variance of U_m by $\phi_m = \mathrm{Var}\,(U_m)$.

Given that the mth observation is taken at level k, let us consider the conditional expectation of U^2_{m+1}.

$$E(U^2_{m+1}/U_m) = (k+1)^2\left(1-\frac{k}{n}\right)+(k-1)^2\frac{k}{n}$$

$$= k^2\left(1-\frac{4}{n}\right)+2k+1.$$

Hence
$$E(U^2_{m+1}) = E(U^2_m)\left(1-\frac{4}{n}\right)+2E(U_m)+1. \tag{5}$$

Now
$$\phi_{m+1} = V(U_{m+1}) = E(U^2_{m+1})-E^2_{m+1},$$

and
$$\phi_m = V(U_m) = E(U^2_m)-E^2_m.$$

Hence

$$\phi_{m+1}+E^2_{m+1} = \phi_m\left[1-\frac{4}{n}\right]+E^2_m\left(1-\frac{4}{n}\right)+2E_m+1. \tag{6}$$

or
$$\phi_{m+1} = \phi_m\left[1-\frac{4}{n}\right]+c_m, \tag{7}$$
where
$$c_m = E_m^2\left(1-\frac{4}{n}\right)+2E_m+1-E_{m+1}^2$$
$$= 1-\frac{4}{n^2}\left(j-\frac{n}{2}\right)^2\left(1-\frac{2}{n}\right)^{2m-2}. \tag{8}$$
Solving (7) as before,
$$\phi_m = A\left(1-\frac{4}{n}\right)^{m-1}+\frac{n}{4}-\left(j-\frac{n}{2}\right)^2\left(1-\frac{2}{n}\right)^{2m-2}, \tag{9}$$
where the arbitrary constant A is determined from the initial condition $\phi_1 = 0$.

Thus
$$0 = A+\frac{n}{4}-\left(j-\frac{n}{2}\right)^2,$$
or
$$A = \left(j-\frac{n}{2}\right)^2-\frac{n}{4}.$$
Hence
$$\phi_m = \left[\left(j-\frac{n}{2}\right)^2-\frac{n}{4}\right]\left(1-\frac{4}{n}\right)^{m-1}+\frac{n}{4}-\left(j-\frac{n}{2}\right)^2\left(1-\frac{2}{n}\right)^{2m-2} \tag{10}$$
is the value of $\mathrm{Var}\,(U_m)$.

4. Small sample up-and-down method (Brownlee, Hodges and Rosenblatt [9])

The probability of response p to stimulus y is given by
$$p = \int_{-\infty}^{(y-u)/\sigma}\frac{1}{\sqrt{2\pi}}e^{-t^2/2}\,dt$$
we fix a system of equally spaced stimuli, $y_0 \pm id$, $i = 0, 1, 2, \ldots$ and perform the first trial at stimulus y_0. Let C denote the sum of the stimuli used in trials $2, 3, \ldots, n+1$, we shall term this the score. As our first estimate for μ we shall consider $\hat{\mu} = C/n$. The asymptotic variance is given in [23]. Let P_0 denote the probability of positive response at stimulus y_0.

We have
$$C_{n+1}(y_0, \mu, d) = \begin{cases} y_0+d+c_n(y_0+d, \mu, d) & \text{if the first trial fails,} \\ y_0-d+c_n(y_0-d, \mu, d) & \text{if the first trial succeeds.}\end{cases}$$

Therefore

$$EC_{n+1}(y_0, \mu, d) = E[c_1(y_0, \mu, d)] + P_0 E[c_n(y_0 - d, \mu, d)] + (1 - P_0) \times E[c_n(y_0 + d, \mu, d)]$$

and

$$\begin{aligned} E[c_{n+1}(y_0, \mu, d) - (n+1)\mu]^2 &= E[y_0 + d - \mu + c_n(v_0 + d, \mu, d) - nu]^2 (1 - P_0) \\ &\quad + E[y_0 - d - \mu + c_n(y_0 - \mu, \mu, d) - nu]^2 P_0 \\ &= E[c_1(y_0, \mu, d) - \mu]^2 + 2(y_0 - \mu)\{E[c_{n+1}(y_0, \mu . d) - (n+1)\mu] \\ &\quad - E[c_1(y_0, \mu, d) - \mu]\} \\ &\quad + 2d\{(1 - P_0) E[c_n(y_0 + d, \mu, d) - n\mu] \\ &\quad - P_0 E - [c_n(y_0 - d, \mu, d) - n\mu]\} \\ &\quad + (1 - P_0) E[c_n(y_0 - d, \mu, d) - n\mu]^2 \\ &\quad + P_0 E[c_n(y_0 - d, \mu, d) - n\mu]^2. \end{aligned}$$

This formulae enable us to compute $[E(\hat{\mu}) - \mu]$ and $E[\hat{\mu} - \mu]^2$. Beginning with $n = 1$ we have

$$E(\hat{\mu}) - \mu = E[c_1(y_0, \mu, d)] - \mu = y_0 - \mu + d(1 - 2P_0)$$

and

$$E(\hat{\mu} - \mu)^2 = E[c_1(y_0, \mu, d) - \mu]^2 = (y_0 - \mu + d)^2 (1 - P_0) + (y_0 - \mu - d)^2 P_0.$$

From these the moments for $n > 1$ follows recursively by using the above formulae.

5. Non-parametric up-and-down method (Derman [20])

Let $y(x)$ be a random variable such that $p(y(x) = 1) = F(x)$ and $P(y(x) = 0) = 1 - F(x)$, where $F(x)$ is a distribution function. It is sometimes of interest, as in sensitivity experiments, to estimate a given quantile of $F(x)$ with observations distributed like $y(x)$ where the choice of x is under control. A procedure for estimating the median is given in §2.2.

A stochastic-approximation process is a general scheme which can be used for estimating any quantile and which imposes no parametric assumptions on $F(x)$. Their method does assume, however, that the range of possible experimental values of x is the real line. In practice, this will not be the case. Limitations such as when x is obtained by a counting procedure, will usually restrict the experimental range of x to a set of numbers of the form

$$a+hn(-\infty < a < \infty, h > 0, n = 0, \pm 1, \ldots).$$

We will consider a non-parametric procedure for estimating any quantile of $F(x)$ on the basis of quantal response data, when experimentally x is restricted to the form $a+hn$.

For the convenience we assume $a = 0$, $h = 1$. Suppose we wish to estimate the value of $x = \theta$ such that

$$F(\theta-0) \leqslant \alpha \leqslant F(\theta) \quad (\tfrac{1}{2} \leqslant \alpha < 1).$$

If $0 < \alpha \leqslant \frac{1}{2}$ or $a \neq 0$ or $h \neq 1$ the necessary modifications are apparent. The experimental procedure is as follows.

Choose x_1 arbitrarily. Recursively, let

$$\begin{aligned} x_n &= x_{n-1}-1 \quad \text{with probability} \quad \frac{1}{2\alpha} && \text{if} \quad y_{n-1} = 1 \\ &= x_{n-1}+1 \quad \text{with probability} \quad 1-\frac{1}{2\alpha} && \text{if} \quad y_{n-1} = 1 \\ &= x_{n-1}+1 \quad \text{with probability} \quad 1 && \text{if} \quad y_{n-1} = 0 \ldots, \end{aligned} \tag{1}$$

where y_k denotes the zero-or-one response at x_k. The estimate θ_n of θ based on n observations is the most frequent value of x, if unique, or the arithmetic average of the most frequent levels, if not unique. We shall prove the following.

THEOREM. *If $F(x)$ is strictly increasing for $\theta-1 \leqslant x \leqslant \theta+1$, then*

$$P(\max[|\limsup_{n\to\infty} \theta_n - \theta|, |\liminf_{n\to\infty} \theta_n - \theta|] < 1) = 1.$$

Let $\{x_n\}$ $(n = 0, 1, \ldots)$ be an irreducible Markov chain with recurrent non-null states and stationary transition probabilities $\{P_{ij}\}$ such that

$$P_{i,i+1}+P_{i,i-1} = 1 \quad (i \ldots 0, \pm 1, \ldots). \tag{2}$$

Let v_i $(i = 0, \pm 1, \ldots)$ be the unique solution of the equation

$$\left\{\begin{array}{ll} \sum_{i=-\infty}^{\infty} v_i p_{ij} = v_j & (j = 0, \pm 1, \ldots), \\ v_i > 0 & \text{for all } i, \\ \sum_{i=-\infty}^{\infty} v_i = 1. & \end{array}\right\} \tag{3}$$

Since $\{x_n\}$ is irreducible and the states are non-null, the system (3) has such a unique solution. The v_i's play the role of stationary absolute probabilities, i.e. if $P(x_0 = i) = v_j$, then $P(x_n = i) = v_i$ for every n.

LEMMA 1. *If for some* $i = b$, $P_{b,b+1} < P_{n,b-1}, P_{b,b+1} > P_{b+1,b+2}$ *and* $P_{i,i+1}$ *is non-increasing in* i *for* $i \geqslant b+1$, *then* $v_b = v_{b+1}$ *and* v_i *is non-increasing in* i *for* $i \geqslant b+1$. *Similarly, if for some* $i = c$, $P_{c,c-1} \leqslant P_{c,c+1}, P_{c\,c-1} > P_{c-1,c-2}$ *and* $P_{i,i+1}$ *is non-decreasing in* i *for* $i \leqslant c-1$, *then* $v_c > v_{c-1}$ *and* v_i *is non-decreasing in* i *for* $i \leqslant c-1$.

Proof. Let $\Pi_{ij} = P(X_n = j$ for some $n \geqslant 1$, $X_r \neq i$ or j for $r < n | X_0 = i)$.

It is known that

$$\frac{v_{i+1}}{v_i} = \frac{\Pi_{i,i+1}}{\Pi_{i+1,i}}. \tag{4}$$

It is clear however that $\Pi_{i,i+1} = P_{i,i+1}$ and $\Pi_{i+1,i} = P_{i+1,i}$. Hence, from (4) and by the hypothesis

$$\frac{v_{b+1}}{v_b} = \frac{P_{b,b+1}}{P_{b+1,b}} = \frac{P_{b,b+1}}{1-P_{b+1,b+2}} < \frac{P_{b,b+1}}{1-P_{b,b+1}} \leqslant 1$$

and thus $v_{b+1} < v_b$. The remainder of the proof follows in the same manner.

Let $N_n(i)$ denote the number of r such that $x_r = i$ for $r \leqslant n$. For the truth of the following lemma we need not impose the condition (2).

LEMMA 2. *Let* B *be the set of states such that* $v_i' = \max\{v_i\}$ *for* $i' \in B$. *Then for every* $i' \in B$,

$$P\left(\lim_{n\to\infty} \frac{N_n(i')}{n} = v_i' > \lim_{n\to\infty} \max_{i \notin B} \left\{\frac{N_n(i)}{n}\right\}\right) = 1.$$

Proof. Since $\sum_{i=-\infty}^{\infty} v_i = 1$, there exists a finite set A of states with $B \subset A$ such that $\sum_{i \notin A} v_i < v_i'$. From the strong law of large numbers of Markov chains, it follows that

$$P\left(\lim_{n\to\infty} \frac{N_n(i)}{n} = v_i\right) = 1 \quad \text{for every } i$$

and more generally

$$P\left(\lim_{n\to\infty}\sum_{i\notin A}\left(\frac{N_n(i)}{n}=\sum_{i\notin A}v_i\right)\right)=1.$$

Let ϵ be any number such that $0<\epsilon<v'_i-\max(\max_{i\in A-B}\{v_i\},\sum_{i\in A}v_i)$ and let E_N denote the event that $(N_n(i')/n>v'_i-\epsilon$ for all $n>N)$. By the previous remark and since $\{E_N\}$ is a monotone sequence, $\lim_{N\to\infty}P(E_N)=P(\lim_{N\to\infty}E_N)=1$. Therefore there exists an N_1 such that $P(N_n(i')/n>v'_i-\epsilon$ for all $n>N_1)>1-\epsilon/3$.

Similarly, since A is finite, there exists an N_2 such that

$$P(\max_{i\in A-B}\{N_n(i)/n\}<v'_i-\epsilon \quad \text{for all} \quad n>N_2)>1-\epsilon/3$$

and an N_3 such that

$$P(\sum_{i\notin A}N_n(i)/n<v'_i-\epsilon \text{ for all } n>N_3)>1-\epsilon/3.$$

Let $N_0=\max(N_1,N_2,N_3)$. Then it follows that

$$P\left\{\left\{\frac{N_n(i')}{n}\right\}>v'_i-\epsilon>\max\left(\max_{i\in A-B}\left\{\frac{N_n(i)}{n}\right\},\ \sum_{i\notin A}\frac{N_n(i)}{n}\right)\right.$$

$$\left.\text{for all } n>N_0\right\}>1-\epsilon.$$ Since $\epsilon>0$ is arbitrary, we have

$$P\left(\lim_{n\to\infty}\frac{N_n(i')}{n}=v'_i>\limsup_{n\to\infty}\max_{n}\max_{i\notin B}(N_n(i)/n\}\right)=1.$$

The last assertion implies that $\lim_n\max_i(N_n(i)/n)$ exists. By a similar argument applied to the finite set B_1 of states which have the second largest v_i's it follows that

$$\limsup_n\max_i(N_n(i)/n)$$

can be replaced by $\lim_{n\to\infty}\max_{i\notin B}(N_n(i)/n)$. The lemma is proved.

Proof. Let $\{X_n\}$ be the Markov chain defined by (1), i.e. let $X_n=i$ if $x_n=i$. The transition probabilities are of the form

$$P_{i,i+1}=1-\frac{F(i)}{2\alpha},\quad P_{i,i-1}=\frac{F(i)}{2\alpha}.$$

The chain is clearly irreducible and the states can be shown to be recurrent non-null.

The numbers $[\theta]+1$ and $[\theta]$ denotes the largest integer less than or equal to θ, can be taken as b and c of lemma 1. From lemma 1 and the condition of strict monotonicity of $F(x)$ for

$$\theta-1 \leqslant x \leqslant \theta+1,$$

it is clear that $[\theta]$, $[\theta]+1$ or both but no other states belong to B of lemma 2. Thus according to lemma 2, the most frequent state, for n large enough, will be $[\theta]+1$, $[\theta]$ or both with probability 1. In any case, the difference between θ and $[\theta]+1$ or $[\theta]$ or the arithmetic average of the two is less than 1. The theorem is therefore proved.

Let us consider an example of Derman's method and see its mode of convergence in this particular case.

6. Illustrative example

Suppose that our density function is

$$\begin{aligned} f(x) &= x \qquad (0 \leqslant x \leqslant \sqrt{2}), \\ &= 0 \qquad \text{elsewhere.} \end{aligned} \tag{1}$$

Our levels will be $i\sqrt{2}/5$, $i = 0, \pm 1, \pm 2, \ldots$ and this study will illustrate only the case where level one is confined to $i\sqrt{2}/5$, where $i = 0, \ldots, 5$ otherwise the level would either decrease or increase until it reached these levels and from there be confined to this set. This is so because outside of this set the probability of a response or non-response is either zero or one: zero below 0 and one above $\sqrt{2}$.

Then

$$\begin{aligned} P_{i,i-1} &= i^2/2 \qquad (i = 0, \ldots, 5\sqrt{2}/5), \\ &= 0 \qquad \text{otherwise;} \end{aligned}$$

$$\begin{aligned} P_{i,i+1} &= 1-i^2/2 \qquad (i = 0, \ldots, 5\sqrt{2}/5), \\ &= 0 \qquad \text{otherwise.} \end{aligned}$$

Using the fact that

$$P_{jk}(n) = \sum_l P_{jl} P_{lk}(n-1), \tag{2}$$

$$P_{jk}(n) = (1-j^2/2)\,P_{j+1,k}(n-1) + j^2/2P_{j-1,k}(n-1)$$

$$= (1-j^2/2)\left[1-\frac{(j+L)^2}{2}P_{j+2,k}(n-2) + \frac{(j+1)^2}{2}P_{j-2,k}(n-2)\right]$$

$$+j^2/2\left[\left(1-\frac{(j-1)^2}{2}\right)P_{j,k}(n-2)+\frac{(j-1)^2}{2}P_{j-2,K}(n-2)\right]. \quad (3)$$

Using (2), (3) may be further extended until it is just a relation in j's since eventually one comes to $n-n$. For $f(x) \equiv x$, the transition probability matrix is, choosing $\alpha = \frac{1}{2}$,

$$P = \begin{pmatrix} 0 & 1 & 0 & 0 & 0 & 0 \\ 0{\cdot}04 & 0 & 0{\cdot}96 & 0 & 0 & 0 \\ 0 & 0{\cdot}16 & 0 & 0{\cdot}84 & 0 & 0 \\ 0 & 0 & 0{\cdot}36 & 0 & 0{\cdot}64 & 0 \\ 0 & 0 & 0 & 0{\cdot}64 & 0 & 0{\cdot}36 \\ 0 & 0 & 0 & 0 & 1 & 0 \end{pmatrix},$$

$$P^2 = \begin{pmatrix} 0{\cdot}04 & 0 & 0{\cdot}96 & 0 & 0 & 0 \\ 0 & 0{\cdot}1936 & 0 & 0{\cdot}8064 & 0 & 0 \\ 0{\cdot}0064 & 0 & 0{\cdot}4560 & 0 & 0{\cdot}5273 & 0 \\ 0 & 0{\cdot}0576 & 0 & 0{\cdot}7120 & 0 & 0{\cdot}2304 \\ 0 & 0 & 0{\cdot}2034 & 0 & 0{\cdot}7696 & 0 \\ 0 & 0 & 0 & 0{\cdot}64 & 0 & 0{\cdot}36 \end{pmatrix},$$

$$P^4 = \begin{pmatrix} 0{\cdot}0139 & 0 & 0{\cdot}47 & 0 & 0{\cdot}5160 & 0 \\ 0 & 0{\cdot}0839 & 0 & 0{\cdot}7303 & 0 & 0{\cdot}1857 \\ 0{\cdot}0031 & 0 & 0{\cdot}3381 & 0 & 0{\cdot}6596 & 0 \\ 0 & 0{\cdot}0522 & 0 & 0{\cdot}7009 & 0 & 0{\cdot}2469 \\ 0{\cdot}0005 & 0 & 0{\cdot}2824 & 0 & 0{\cdot}7161 & 0 \\ 0 & 0{\cdot}0368 & 0 & 0{\cdot}6861 & 0 & 0{\cdot}2770 \end{pmatrix},$$

$$P^8 = \begin{pmatrix} 0{\cdot}0019 & 0 & 0{\cdot}3112 & 0 & 0{\cdot}6868 & 0 \\ 0 & 0{\cdot}0520 & 0 & 0{\cdot}7005 & 0 & 0{\cdot}2473 \\ 0{\cdot}0001 & 0 & 0{\cdot}3020 & 0 & 0{\cdot}6969 & 0 \\ 0 & 0{\cdot}501 & 0 & 0{\cdot}6987 & 0 & 0{\cdot}4511 \\ 0{\cdot}0012 & 0 & 0{\cdot}2979 & 0 & 0{\cdot}6993 & 0 \\ 0 & 0{\cdot}0490 & 0 & 0{\cdot}6978 & 0 & 0{\cdot}2530 \end{pmatrix},$$

$$P^{16} = \begin{pmatrix} 0\cdot0008 & 0 & 0\cdot2992 & 0 & 0\cdot6985 & 0 \\ 0 & 0\cdot050 & 0 & 0\cdot6984 & 0 & 0\cdot2513 \\ 0\cdot0008 & 0 & 0\cdot2988 & 0 & 0\cdot6978 & 0 \\ 0 & 0\cdot050 & 0 & 0\cdot6985 & 0 & 0\cdot2513 \\ 0\cdot0008 & 0 & 0\cdot2987 & 0 & 0\cdot6975 & 0 \\ 0 & 0\cdot050 & 0 & 0\cdot6990 & 0 & 0\cdot2513 \end{pmatrix}.$$

For $\alpha = \frac{1}{2}$,

$$\theta = \mu = \int_0^{\sqrt{2}} x \, . \, x \, dx = \tfrac{2}{3}\sqrt{2}.$$

Since in any experiment, one would have, on the basis of experience, a disposition towards choosing a certain level, suppose that $i = 2, 3, 4$ are preferred with probabilities $\frac{1}{3}$, $\frac{1}{3}$, $\frac{1}{3}$, of being chosen. Then the probability of ending in each state after 16 experiments is

$$\begin{aligned} pP^{16} &= [0, 0, \tfrac{1}{3}, \tfrac{1}{3}, \tfrac{1}{3}, 0]\, p^{16} \\ &= [0\cdot0006,\ 0\cdot050,\ 0\cdot199,\ 0\cdot233,\ 0\cdot466,\ 0\cdot084] \end{aligned}$$

already the

$$P\left[\left|\max\left(|\sup(\theta_n - \theta)|,\ |\inf(\theta_n - \theta)|\right)\right| < \frac{\sqrt{2}}{6}\right] \approx 0\cdot7.$$

In this case the convergence is quite rapid.

7. Problems

1. Assume the set up of §3 and evaluate the covariance of U_m and U_{m+1}. (Narayana)

2. Assume the set up of §5. Show that $\{X_n\}$ is recurrent.

APPENDIX 1

ITERATIVE TECHNIQUES OF NUMERICAL ANALYSIS

1. Introduction

Iterative techniques have been well-known mathematical tools since before the days of Newton. They have utility in both theoretical and applied work. Furthermore their utility has been enhanced by the invention of the computer which can do in seconds a calculation which once took days. And these techniques are frequently employed in problems such as that of locating a point where a given function behaves in a prescribed way, as for example when one is interested in finding the root of an equation, or locating a point of maximum of a function. Here we discuss two methods of iteration, one called the Pre-Newton–Raphson method and the other called the Newton–Raphson method. Then we study the conditions under which the sequences generated by these methods result in convergence to the desired point of solution. For the details of this appendix and related topics one can refer to Ostrowski [61] and Traub [70].

2. Pre-Newton–Raphson method

Let M be a function from the real number interval $I = [a, b]$ to I. One is interested in solving the following equation

$$M(x) = \alpha, \tag{1}$$

where the functional form of M is not known, but it is known that M is increasing and continuous. This can be done by the following method. Choose X_1 arbitrarily from I and consider the sequence of $\{X_n\}$ given by (2).

$$X_{n+1} = X_n - a^{-1}[M(X_n) - \alpha], \quad \text{where} \quad a \neq 0. \tag{2}$$

Then under suitable conditions which we describe in §4 $\{X_n\}$ converges to θ such that $M(\theta) = \alpha$. This procedure is analogous to method of proportional control. We illustrate this method by figure 1. A natural question is the choice of a. One can see that in order to have high rate of convergence one should have $a = M'(\theta)$.

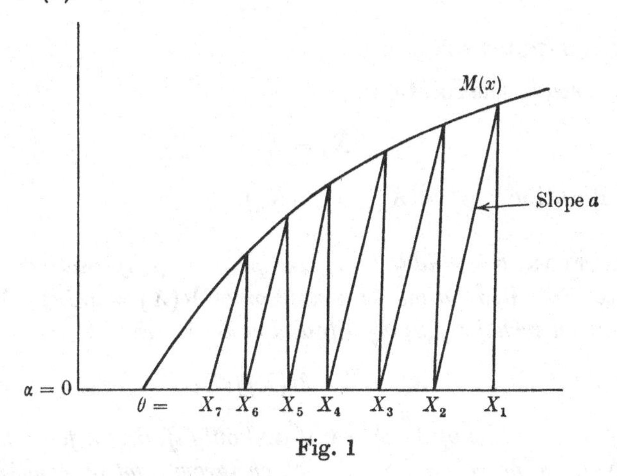

Fig. 1

Since θ is unknown this leads to the following method.

3. Newton–Raphson method

Let us assume the set up of the previous section and furthermore suppose that the derivative at each stage of iteration can be computed. Then one can define a recursive scheme as follows.

Choose X_1 arbitrarily in the interval I and

$$X_{n+1} = X_n - a_n^{-1}[M(X_n) - \alpha], \tag{1}$$

where

$$a_n = \begin{cases} a_1 & \text{if} \quad M'(X_n) < a_1, \\ M'(X_n) & \text{if} \quad a_1 \leqslant M'(X_n) \leqslant a_2, \\ a_2 & \text{if} \quad M(X_n) > a_2, \end{cases} \tag{2}$$

where a_1 and a_2 are constants such that $0 < a_1 < a_2 < \infty$. This method of iteration is called the Newton–Raphson method. Though this method involves the computation of the derivative of M at each stage of iteration, the eventual rate of convergence is improved. For further discussion one can refer to Ostrowski

[61]. Now we take up the problem of existence of the solution and its uniqueness. This is done usually for a fixed-point problem, which implies a result for the problem under consideration.

4. A fixed-point problem

Let us study the solution of

$$\psi(X) = X \tag{1}$$

by the iteration

$$X_{n+1} = \psi(X_n). \tag{2}$$

DEFINITION. *θ is said to be a fixed point of ψ, if θ satisfies* (1). *Let us consider its relation to the solution of $M(X) = 0$. Let $f(X)$ be any function such that $f(\theta)$ is finite and non-zero, and let*

$$\psi(X) = X - M(X)f(X). \tag{3}$$

Then θ is a solution of $M(x) = 0$ if and only if, θ is a fixed point of ψ. Now we prove some results which throw light on conditions under which a solution of the fixed-point problems exists.

LEMMA 1. *Let ψ be a continuous function from the bounded and closed interval $I = [a, b]$ to I Then there exists θ, $a \leqslant \theta \leqslant b$ such that $\psi(\theta) = \theta$.*

Proof. Since the function is from I to I

$$\psi(a) \geqslant a, \quad \psi(b) \leqslant b.$$

Let $\phi(x) = \psi(x) - x$. Hence $\phi(a) \geqslant 0$ and $\phi(b) \leqslant 0$, then there exists θ such that $\phi(\theta) = 0$, that is $\psi(\theta) = \theta$ as desired.

Lipschitz condition.

In order to draw additional conclusions, we have to impose additional restriction on ψ. Let

$$|\psi(c) - \psi(d)| \leqslant K|c-d|, \quad 0 \leqslant K < 1 \tag{4}$$

for arbitrary points c and d in I. It is easy to see that condition (4) implies the continuity of ψ. Now we prove uniqueness of solution of $\psi(x) = x$.

LEMMA 2. *Let ψ be a function from I to I which satisfies* (4). *Then $\psi(x) = x$ has at most one solution.*

Proof. Let us assume that there are two distinct solutions, θ_1 and θ_2. Then

$$\theta_1 = \psi(\theta_1), \quad \theta_2 = \psi(\theta_2),$$

$$|\theta_1 - \theta_2| = |\psi(\theta_1) - \psi(\theta_2)| \leqslant K|\theta_1 - \theta_2| < |\theta_1 - \theta_2|$$

which is not true. Hence $\theta_1 = \theta_2$.

REMARK. We have established the conditions for existence and uniqueness of a solution θ of the equation $\psi(x) = x$. Now we attempt to show that the sequence $x_{i+1} = \psi(x_i)$ converges to this solution.

LEMMA 3. *Let I be a closed, bounded interval and let ψ be a function from I to I which satisfies* (4). *Let X_0 be an arbitrary point of I. Let $X_{i+1} = \psi(X_i)$. Then the sequence X_i converges to the unique solution of $\psi(x) = X$ in I.*

Proof.
$$X_{i+1} - \theta = \psi(X_i) - \theta = \psi(X_i) - \psi(\theta).$$

From (4) we can conclude that

$$|X_{i+1} - \theta| = |\psi(X_i) - \psi(\theta)| \leqslant K|X_i - \theta|,$$

where $k < 1$.

Thus by iteration we can obtain

$$|X_{i+1} - \theta| \leqslant K^i|X_0 - \theta|,$$

when $i \to \infty, K^i \to 0$. Because $0 \leqslant K < 1$. Hence $X_i \to \theta$ as $i \to \infty$.

An estimate of error.

One can make an estimate of the error of rth approximant, which depends only on the first two approximants and the Lipschitz constant. From condition (4) we obtain for every i

$$|X_{i+1} - X_i| \leqslant K^i|X_1 - X_0|. \tag{5}$$

Let r and p be arbitrary positive integers. Then

$$X_{r+p} - X_r = (X_{r+p} - X_{r+p-1}) + (X_{r+p-1} - X_{r+p-2}) + \dots + (X_{r+1} - X_r),$$

and using the fact that modulus of the sum is less than the sum of moduli

$$|X_{r+p}-X_r| \leqslant |X_{r+p}-X_{r+p-1}| + |X_{r+p-1}-X_{r+p-2}|+\ldots+|X_{r+1}-X_r|. \quad (6)$$

Then by applying (5) we obtain

$$|X_{r+p}-X_r| \leqslant K^r(1+K+\ldots+K^{p-1})|X_1-X_0| = \frac{K^r(1-K^p)}{1-K}|X_1-X_0|. \quad (7)$$

Then $$|X_r-\theta| \leqslant \frac{K^r}{1-K}|X_1-X_0| \quad \text{as} \quad p\to\infty. \quad (8)$$

APPENDIX 2

LIMIT THEOREMS

1. Introduction

In this appendix those limit theorems are discussed which are the basis of stochastic-approximation techniques. In §2 we study convergence of sequences of random variables. One can find most of these results in Loevé [58].

In §3 we discuss properties of multidimensional characteristic function which are extensions of the result to be found in Loevé [58], and in particular a result of Sacks [67] which enables him to give a simplified proof of asymptotic normality. Finally, some theorems on conditional expectations are studied (with their modification) which one can find in Doob [22] and Loevé [58].

2. Convergence of sequence of random variables

THEOREM 1. *A necessary and sufficient condition that the random variable sequence $\{X_n\}$ converges to the random variable X with probability one is that if each ϵ and η is a positive number then there is a positive integer m_0 such that if $m_0 \leqslant m \leqslant n$ then*

$$p[\sup_{m \leqslant i \leqslant n} |X_i - X| \geqslant \epsilon] < \eta.$$

Proof. We note that the condition is equivalent to

$$\lim_{m\to\infty} \lim_{n\to\infty} p\left[\sup_{m \leqslant i \leqslant n} |X_i - X| \geqslant \frac{1}{k}\right] = 0$$

for each positive integer k. Since

$$\left[\sup_{m \leqslant i \leqslant n} |X_i - X| \geqslant \frac{1}{k}\right] = \bigcup_{i=m}^{n} \left[|X_i - X| \geqslant \frac{1}{k}\right],$$

the condition is equivalent to

$$\lim_{m\to\infty} \lim_{n\to\infty} p\left\{\bigcup_{i=m}^{n} \left[|X_i - X| \geqslant \frac{1}{k}\right]\right\} = 0.$$

We notice that

$$\lim_{m\to\infty}\lim_{n\to\infty} p\left\{\bigcup_{i=m}^{n}\left[|X_i - X| \geqslant \frac{1}{k}\right]\right\}$$

$$= \lim_{m\to\infty} p\left\{\bigcup_{i=m}^{\infty}\left[|X_i - X| \geqslant \frac{1}{k}\right]\right\}$$

$$= \lim_{m\to\infty} p\left\{\bigcap_{j=1}^{m}\bigcup_{i=j}^{\infty}\left[|X_i - X| \geqslant \frac{1}{k}\right]\right\}$$

$$= p\left\{\bigcap_{j=1}^{\infty}\bigcup_{i=j}^{\infty}\left[|X_i - X| \geqslant \frac{1}{k}\right]\right\}.$$

Now $\quad p[\lim_{n\to\infty} X_n = X] = p\left\{\bigcap_{k}\bigcup_{j}\bigcup_{i\geqslant j}\left[|X_i - X| < \frac{1}{k}\right]\right\},$

therefore

$$p[\lim_{n\to\infty} X_n \neq X] = p\left\{\bigcup_{k}\bigcap_{j}\bigcup_{i\geqslant j}\left[|X_i - X| \geqslant \frac{1}{k}\right]\right\}$$

and

$$p\left\{\bigcap_{j}\bigcup_{i\geqslant j}\left[|X_i - X| \geqslant \frac{1}{k}\right]\right\} \leqslant p\left\{\bigcup_{k}\bigcap_{j}\bigcup_{i\geqslant j}\left[|X_i - X| \geqslant \frac{1}{k}\right]\right\}$$

$$\leqslant \sum_{k=1}^{\infty} p\left\{\bigcap_{j}\bigcup_{i\geqslant j}\left[|X_i - X| \geqslant \frac{1}{k}\right]\right\}.$$

If the condition of the theorem holds, then the right-hand side is zero, implying $p\{\lim_{n\to\infty} X_n \neq X\} = 0$. If on the other hand

$$p[\lim_{n\to\infty} X_n \neq X] = 0, \quad \text{then} \quad p\left\{\bigcap_{j}\bigcup_{i\geqslant j}\left[|X_i - X| \geqslant \frac{1}{k}\right]\right\} = 0$$

for all k, implying the condition.

THEOREM 2. *A necessary and sufficient condition that the random variable sequence $\{X_n\}$ converges to some random variable with probability one is that if each of ϵ and η is a positive number then there is a positive integer m_0 such that if $m_0 \leqslant m < n$, then*

$$p[\sup_{m\leqslant i<j\leqslant n} |X_i - X_j| \geqslant \epsilon] < \eta.$$

Proof. The condition is equivalent to

$$\lim_{m\to\infty}\lim_{n\to\infty} p\left(\bigcup_{i=m}^{n}\bigcup_{j=m}^{n}\left[|X_i - X_j| \geqslant \frac{1}{k}\right]\right) = 0$$

for each positive integer k. We have that

$$\lim_{m\to\infty}\lim_{n\to\infty} p\left(\bigcup_{i=m}^{n}\bigcup_{j=m}^{n}\left[|X_i-X_j|\geqslant\frac{1}{k}\right]\right)$$

$$= \lim_{m\to\infty} p\left\{\bigcup_{i=m}^{\infty}\bigcup_{j=m}^{\infty}\left[|X_i-X_j|\geqslant\frac{1}{k}\right]\right\}$$

$$= \lim_{m\to\infty} p\left(\bigcap_{r=1}^{m}\bigcup_{i=r}^{\infty}\bigcup_{j=r}^{\infty}\left[|X_i-X_j|\geqslant\frac{1}{k}\right]\right)$$

$$= p\left\{\bigcap_{r=1}^{\infty}\bigcup_{i=r}^{\infty}\bigcup_{j=r}^{\infty}\left[|X_i-X_j|\geqslant\frac{1}{k}\right]\right\}$$

$$= p\left\{\bigcap_{m=1}^{\infty}\bigcup_{i\geqslant m}^{\infty}\bigcup_{j\geqslant m}^{\infty}\left[|X_i-X_j|\geqslant\frac{1}{k}\right]\right\}.$$

Since $\{X_n\}$ is a Cauchy sequence, the set

$$\left\{\bigcap_{k}\bigcup_{m}\bigcap_{i\geqslant m}\bigcap_{j\geqslant m}\left[|X_i-X_j|<\frac{1}{k}\right]\right\}\in F, \text{ the Borel field.}$$

Hence its complement set

$$\left\{\bigcup_{k}^{\infty}\bigcap_{m=1}^{\infty}\bigcup_{i\geqslant m}^{\infty}\bigcup_{j\geqslant m}^{\infty}\left[|X_i-X_j|\geqslant\frac{1}{k}\right]\right\}\in F$$

is also in F.

Thus we have

$$p\left\{\bigcap_{m}^{\infty}\bigcup_{i\geqslant m}^{\infty}\bigcup_{j\geqslant m}^{\infty}\left[|X_i-X_j|\geqslant\frac{1}{k}\right]\right\}$$

$$\leqslant p\left\{\bigcup_{k}^{\infty}\bigcap_{m}^{\infty}\bigcup_{i\geqslant m}^{\infty}\bigcup_{j\geqslant m}^{\infty}\left[|X_i-X_j|\geqslant\frac{1}{k}\right]\right\}$$

$$\leqslant \sum_{k=1}^{\infty} p\left\{\bigcap_{m}^{\infty}\bigcup_{i\geqslant m}^{\infty}\bigcup_{j\geqslant m}^{\infty}\left[|X_i-X_j|\geqslant\frac{1}{k}\right]\right\}.$$

If the condition of the theorem holds, then the right-hand side is zero. If the conclusion holds, then the left-hand side is zero. Now if $\{X_n\}$ is a Cauchy sequence then

$$[-\infty < \liminf_{n\to\infty} X_n = \limsup_{n\to\infty} X_n < \infty].$$

Note

$$\liminf_{n\to\infty} X_n = \sup_{k}\inf_{n\geqslant k} X_n$$

$$= \limsup_{n\to\infty} X_n$$

$$= \inf_{k}\sup_{n\geqslant k} X_n.$$

Now
$$[\inf_{n\geqslant k} X_n < X] = \bigcup_{n\geqslant k} [X_n < X]$$
implying that $\inf_{n\geqslant k} X_n$ is a measurable function (random variable is finite valued). If $\liminf_{n\to\infty} X_n = \limsup_{n\to\infty} X_n$ is finite valued, then $\{X_m\}$ has a limit with probability 1, and this limit is a random variable.

BOREL–CANTELLI LEMMA. *If each of $A_1, A_2, \ldots, A_n$ is in F and $\sum_n p(A_n) < \infty$ then*
$$p\left(\bigcap_m^{\infty} \bigcup_{n\geqslant m} A_n\right) = 0,$$
$$\bigcap_m \bigcup_{n\geqslant m} A_n = \limsup_{n\to\infty} A_n$$
$$= \{\omega|\omega \textit{ belongs to an infinite number of } A_n\text{'s}\}$$
$$p\left(\bigcup_m^{\infty} \bigcup_{n\geqslant m}^{\infty} A_n\right) = \lim_{m\to\infty} \lim_{n\to\infty} p\left(\bigcup_{j=m}^{n} A_j\right). \tag{1}$$
Since $\sum_n p(A_n) < \infty$, (1) will go to zero as $n, m \to \infty$.

THEOREM 3. *If for some $r > 0$, $\sum_n E|X_n - X|^r < \infty$, then $X_n \to X$ with probability one.*

Proof.
$$\sum_n p[|X_n - X| \geqslant \epsilon] \leqslant \epsilon^{-r} \sum_n^{\infty} E|X_n - X|^r < \infty.$$
Hence the condition of Theorem 1 is satisfied.

COROLLARY 2·1. *A necessary and sufficient condition that the random variable sequence $\{X_n\}$ converge to some random variable with probability one is that if each of ϵ and η is a positive number then there is a positive integer m_0 such that if $n > m \geqslant m_0$ then*
$$p[\sup_{m\leqslant j\leqslant n} |X_j - X_m| \geqslant \epsilon] < \eta.$$

Proof. This follows from the fact that
$$p[\sup_{m\leqslant j\leqslant n} |X_j - X_m| \geqslant \epsilon] \leqslant p[\sup_{m\leqslant i<j\leqslant n} |X_i - X_j| \geqslant \epsilon]$$
$$\leqslant p[\sup_{m\leqslant j\leqslant n} |X_j - X_m| \geqslant \epsilon/2].$$

Because
$$A = [\omega|\sup_{m\leqslant j\leqslant n}|X_j - X_m| \geqslant \epsilon],$$

$$B = [\omega|\sup_{m\leqslant i\leqslant j\leqslant n}|X_i - X_j| \geqslant \epsilon],$$

$$C = [\omega|\sup_{m\leqslant j\leqslant n}|X_j - X_n| \geqslant \epsilon/2]$$

and $A \subset B \subset C$.

Suppose a point $\omega \in B$, then there is a pair (i,j) such that $m \leqslant i < j \leqslant n$ and $|X_i(\omega) - X_j(\omega)| \geqslant \epsilon$. This implies that either

$$|X_i(\omega) - X_m(\omega)| \geqslant \epsilon/2 \quad \text{or} \quad |X_j(\omega) - X_m(\omega)| \geqslant \epsilon/2,$$

since otherwise

$$|X_i(\omega) - X_j(\omega)| \leqslant |X_i(\omega) - X_m(\omega)| + |X_j(\omega) - X_m(\omega)| < \epsilon/2 + \epsilon/2 = \epsilon.$$

Then if $\omega \in B$, $\omega \in C$.

Hence the result follows from the Theorem 2.

LEMMA 1 (Burkholder). *Suppose F_n is continuously convergent at θ to β (that is, if $\{y_n\}$ is a real number sequence with limit θ, then $F_n(y_n) \to \beta$), then if $\epsilon > 0$ there is a $\delta > 0$ and an m such that if $|x-\theta| < \delta$ and $n > m$ then $|F_n(x) - \beta| < \epsilon$.*

Proof. Suppose the above is not true, then there is an $\epsilon > 0$ such that for all pairs (δ, m) where $\delta > 0$ and m is a positive integer there is an (x, n) satisfying $|x-\theta| < \delta$, $n > m$ and $|F_n(x) - \beta| \geqslant \epsilon$. Thus for all n and k's there is a pair (x_k, n_k) such that $|x_k - \theta| < (1/k)$, $n_k > k$ and $|F_{n_k}(x_k) - \beta| \geqslant \epsilon$. Let $\{y_n\}$ be any sequence with limit θ such that $y_{n_k} = x_k$, then

$$|F_{n_k}(y_{n_k}) - \beta| \geqslant \epsilon.$$

But $\lim_{n\to\infty} F_n(y_n) = \beta$, therefore $\lim_{n\to\infty} F_n(y_{n_k}) = \beta$, which is a contradiction to the supposition.

LEMMA 2 (Burkholder). *Let $\{X_n\}$ be a random variable sequence and if n is a positive integer let each of F_n and g_n be a real-valued Borel measurable function from the real numbers. Suppose $\{F_n\}$ is continuously convergent at the number θ to the number β and is* *uniformly bounded. Suppose that if n is in N then* $E[g_n(X_n)]$ *exists finitely and if* $\delta > 0$,

$$E\{|g_n(X_n)| \,|\, |X_n-\theta| \geqslant \delta\} p\{|X_n-\theta| \geqslant \delta\} = o(1)\, E|g_n(X_n)|.$$

Then

$$E\{F_n(X_n)\, g_n(X_n)\} = \beta E[g_n(X_n)] + o(1)\, E|g_n(X_n)|.$$

Proof. Let $K = \sup_{n,x} F_n(x)$. By assumption K is finite. The assumption also implies that $E[F_n(X_n)\, g_n(X_n)]$ exists finitely for all n. Let $\epsilon > 0$ and since $\{F_n\}$ is continuously convergent at θ to β there is a $\delta > 0$ and an m in N such that if $|x-\theta| < \delta$ and $n > m$ then $|F_n(x)-\beta| < \epsilon$. Thus for $n > m$.

$$\begin{aligned} |E\{(F_n(X_n)-\beta)\, g_n(X_n)\}| &\leqslant E\{|F_n(X_n)-\beta|\,|g_n(X_n)|\} \\ &\leqslant \epsilon E\{|g_n(X_n)| \,|\, |X_n-\theta| < \delta\} p\{|X_n-\theta| \leqslant \delta\} \\ &\quad + (K+|\beta|)\, E\{|g_n(X_n)| \,|\, |X_n-\theta| \geqslant \delta\} p\{|X_n-\theta| \geqslant \delta\} \\ &\leqslant [\epsilon + o(1)]\, E|g_n(X_n)| \end{aligned}$$

which implies the desired result.

LEMMA 3. *If X and Y are random variables and* ϵ *is a positive number such that* $p[X \neq Y] < \epsilon$, *then* $|F(x)-G(x)| < \epsilon$ *for all real* x, *where F is the distribution function of X and G of Y.*

Proof. For each real x,

$$\begin{aligned} F(x) &= p[X \leqslant x] \\ &= p[X \leqslant x | X = Y]\, p[X = Y] \\ &\qquad + p[X \leqslant x | X \neq Y] \times p[X \neq Y] \\ &= p[Y \leqslant x | X = Y]\, p[X = Y] \\ &\qquad + p[Y \leqslant x | X \neq Y]\, p[X \neq Y] \\ &\qquad - p[Y \leqslant x | X \neq Y]\, p[X \neq Y] \\ &\qquad + p[X \leqslant x | X \neq Y] \times p[X \neq Y] \\ &< p[Y \leqslant x] + \epsilon = G(x) + \epsilon. \end{aligned}$$

Therefore $[F(x)-G(x)] < \epsilon$. Similarly one can show that

$$[G(x)-F(x)] < \epsilon, \quad \text{hence} \quad |F(x)-G(x)| < \epsilon.$$

LEMMA 4 (Burkholder). *If $G_n(y|.)$ is Borel measurable, then R_n is Borel measurable where*

$$R_n(x) = \int_{-\infty}^{\infty} y\, dG_n(y|x).$$

Proof. If for each positive integer m, $j_m(x) = \int_{-m}^{m} y\, dG(y|x)$ is Borel measurable in x then R_n is, since $j_m(x) \to R_n(x)$ for each real x.

Let

$$S_k(x) = \sum_{j=1}^{k} \left(-m+\frac{j}{k}\right)\left[G_n\left(-m+\frac{j}{k}\Big/x\right) - G_n\left(-m+\frac{j-1}{k}\Big/x\right)\right]$$

then $S_k(x) \to j_m(x)$ as $k \to \infty$, since $|j_m(x)-S_k(x)| \leqslant 1/k$. But S_k is Borel measurable, thus result follows.

DEFINITION. *If A is a real number set, f is a complex-valued function whose range of definition includes A, and b is a positive number, let*

$$C(A,f,b) = \sup\{|f(x)-f(y)|\,|x \text{ in } A, Y \text{ in } A, |x-y| \leqslant b\}.$$

DEFINITION. *If g is a function from a real number interval I into R, g will be called Borel measurable on I if the function f is Borel measurable where*

$$\begin{aligned} f(x) &= g(x) \quad \text{if} \quad x \in I \\ &= 0 \quad\quad\; \text{if} \quad x \notin I. \end{aligned}$$

THEOREM 4 (Burkholder). *Let I be a real number interval and for each x in I let $G(./x)$ be a distribution function such that if Y is in R then $(G(Y/.)$ is Borel measurable on I. Let f be a function from $R \times I$ such that if x is in I then $\int_{-\infty}^{\infty} f(Y,x)\, dG(Y/x)$ exists, if y is in R then $f(Y,\cdot)$ is Borel measurable on I and such that for each m in*

N, $C((-m,m), f(\cdot,x), b)\to 0$ uniformly for x in I as $b\to 0$. Then the function J is Borel measurable on I, where for x in I

$$J(x) = \int_{-\infty}^{\infty} f(Y,x)\,dG(Y|x).$$

Proof. If for each m in N the function J_m, where for x in I $J_m(x) = \int_{-m}^{m} f(y,x)\,dG(y/x)$, is Borel measurable on I then J is Borel measurable on I since $\{J_m\}$ converges to J.

Let m be in N. The assumptions imply that $f(\cdot,x)$ is continuous for each x in I. Therefore, $\int_{-m}^{m} f(y,x)\,dG(y/x)$ is the Stieltjes integral of $f(\cdot,x)$ with respect to $G(.|x)$ since, of course $G(./x)$ is of bounded variation. Thus, there is a natural number sequence $\{k_n\}$ and a double sequence $\{t_{ij}\}$ $i = 1,2,\ldots,$; $j = 1,\ldots,k_n+1$; such that

$$-m = t_{11} < t_{12} < \ldots < t_{1,k_n+1} = m$$

$$\ldots\ldots\ldots\ldots\ldots\ldots\ldots\ldots\ldots\ldots\ldots\ldots$$

$$-m = t_{n1} < t_{n2} < \ldots < t_{n,k_n+1} = m$$

$$\ldots\ldots\ldots\ldots\ldots\ldots\ldots\ldots\ldots\ldots\ldots\ldots$$

$$\max_{1\leqslant i\leqslant k_n} (t_{n,i+1}-t_{n,i}) < 1/n,$$

and for each x in I

$$\left|\int_{-m}^{m} f(y,x)\,dG(y/x) - S_n(x)\right| \leqslant \sup_{x \text{ in } I} C([-m,m], f(\cdot,x), 1/n),$$

where $$S_n(x) = \sum_{i=1}^{n} f(t_i,x)\,[G(t_{i+1}/x) - G(t_i/x)].$$

By the assumptions, $\{S_n\}$ converges to J_m on I. Clearly, S_n is Borel measurable on I for n in N. Therefore J_n is Borel measurable on I. This completes the proof of the theorem.

THEOREM 5 (Burkholder). *For each x in R let $(Y_1(x)\ldots Y_n(x))$ be an independent random variable set where $Y_k(x)$ has the distribution function $H_k(./x)$, $k = 1,\ldots,n$ and $G(./x)$ is the distribution function of $\sum_{k=1}^{n} Y_k(x)$. Suppose that if y is in R then $H_k(y/.)$ is Borel measurable, $k = 1,\ldots,n$. Then for each y in R, $G(y/.)$ is Borel measurable.*

Proof. Let $\phi_k(.|x)$ be the characteristic function of $Y_k(x)$ $k = 1, \ldots, m$, and let $\phi(.|x)$ be the characteristic function of $\sum_{k=1}^{n} Y_k(x)$. Thus, if (t, x) is in R^2 then

$$\phi(t|x) = \prod_{k=1}^{n} \phi_k(t|x). \tag{1}$$

Since

$$\phi_k(t/x) = \int_{-\infty}^{\infty} e^{ity} dH_k(y/x)$$

$$= \int_{-\infty}^{\infty} \cos ty \, dH_k(y/x) + i \int_{-\infty}^{\infty} \sin ty \, dH_k(y/x),$$

applying the Theorem 4 gives that if t is in R then each of

$$\operatorname{Re} \phi_k(t/.) \quad \text{and} \quad I_m \phi_k(t/.)$$

is Borel measurable, $k = 1, 2, \ldots, n$. A similar statement holds for $\phi(t)$ using (1). From the relation

$$|\phi(t_1/x) - \phi(t_2/x)| \leqslant \int_{-\infty}^{\infty} |e^{it_1 y} - e^{it_2 y}| dG(y/x)$$

$$\leqslant \sup_{y} |e^{it_1 y} - e^{it_2 y}|$$

it is evident that $\quad C(R, \phi(\cdot/x), b) \to 0$

uniformly in x as $b \to 0$. By Robbins' variation of the Lévy inversion formula we have that if (y, x) is in R^2 then

$$2G(y/x) = G(y^-|x) + G(y^+|x)$$

$$= 1 - \lim_{T \to \infty} \left(\int_{-T}^{-1/T} + \int_{1/T}^{T} \right) e^{-ity} \frac{\phi(t/x)}{t\pi i} dt.$$

Using methods similar to those used in the proof of the Theorem 4 it can be shown that if y is in R then $\bar{G}(y/\cdot)$ is Borel measurable. We now show that this implies the desired result.

Let y be in R. For each n in

$$N, \bar{G}\left(y + \frac{1}{n} \middle/ .\right)$$

is Borel measurable. Thus, since

$$\bar{G}\left(y+\frac{1}{n}\middle/\cdot\right) \to G(y/\cdot) \quad \text{as} \quad n \to \infty, \quad G(y/.)$$

is Borel measurable.

3. Multidimensional characteristic functions

With all vectors considered as elements of q-dimensional Euclidean space we adopt the following notation. If $\mathbf{x}, \mathbf{y}$ are vectors $\langle \mathbf{x}, \mathbf{y} \rangle$ will denote their inner product. The norm of a vector $\mathbf{x}$ we denote by $\|\mathbf{x}\|$ and, of course, is equal to $\langle \mathbf{x}, \mathbf{x} \rangle^{\frac{1}{2}}$. If A is a $q \times q$ matrix we define in the usual way,

$$\|A\|_s = \sup_{\|\mathbf{x}\|=1} \langle A\mathbf{x}, A\mathbf{x} \rangle^{\frac{1}{2}}.$$

The obvious facts that $\|A\mathbf{x}\| \leqslant \|A\|_s \|\mathbf{x}\|$ and that

$$\|A_1 A_2\|_s < \|A_1\|_s \|A_2\|_s$$

will be useful. I will denote the identity $q \times q$ matrix A' and x' will denote the transposes of the matrix A and vector $\mathbf{x}$ respectively. Unless otherwise indicated a vector is to be considered a column vector.

Let $\{U_{nk};\ 1 \leqslant k \leqslant n, n \geqslant 1\}$ be a family of vector random variables, the distribution of U_{nk} being denoted by F_{nk}. Let

$$V_{nk} = (U_{n_1}, \ldots, U_{n_{k-1}})$$

and suppose that $E(U_{kn} | V_{nk}) = 0$ with probability one. Denote the covariance matrix of U_{nk} by S_{nk}, i.e. $S_{nk} = E(U_{nk} U'_{nk})$. Let $r_{nk} - E(U_{nk} U'_{nk} | V_{nk})$. Let $U_n = \Sigma U_{nk}$, $S_n = \Sigma S_{nk}$ and $r_n = \Sigma r_{nk}$, where all three summations are over $1 \leqslant k \leqslant n$. For $\epsilon > 0$ define

$$\begin{aligned} \phi^{\epsilon}_{nk} &= 1 \quad \text{if} \quad \|U_{nk}\| > \epsilon \\ &= 0 \quad \text{otherwise.} \end{aligned}$$

THEOREM 6 (Sacks). *If*

$$\lim_{n \to \infty} \sum_{k=1}^{n} E(\|r_{nk} - S_{nk}\|_s) = 0, \tag{1}$$

$$\sup_{n} \sum_{k=1}^{n} E(\|U_{nk}\|^2) < \infty \tag{2}$$

and, for every $\epsilon > 0$

$$\lim_{n\to\infty} \sum_{k=1}^{n} E(\|U_{nk}\|^2 \phi^{\epsilon}_{nk}) = 0 \tag{3}$$

and $S_n \to S$, *i.e.* $\|S_n - S\|_s \to 0$ *then* U_n *is asymptotically normal with mean zero and covariance matrix* S.

Proof. Let F and G be q-dimensional distribution functions with characteristic functions f and g and finite covariance matrices C and D respectively, and let $H = F - G$. Let θ_1 and θ_2 denote quantities whose absolute value is less than 1. Let $\Delta = \{\mathbf{x} \mid \|\mathbf{x}\| \leqslant \epsilon\}$ and let Δ' be the complement of Δ. Then, for fixed $\mathbf{t}$ and

$$\epsilon > \frac{1}{\|t\|}$$

$$\begin{aligned}
|f(\mathbf{t}) - g(\mathbf{t})| &\leqslant \left|\int \langle \mathbf{t}, \mathbf{x}\rangle \, dH(\mathbf{x})\right| + \frac{1}{2}\left|\int_{\Delta} \langle \mathbf{t}, \mathbf{x}\rangle^2 \, dH(\mathbf{x})\right| \\
&\quad + \left|\int_{\Delta} \theta_1 \langle \mathbf{t}, \mathbf{x}\rangle^3 \, dH(\mathbf{x})\right| + \left|\int_{\Delta'} \theta_2 \langle \mathbf{t}, \mathbf{x}\rangle^2 \, dH(\mathbf{x})\right| \\
&\leqslant \left|\int \langle \mathbf{t}, \mathbf{x}\rangle \, dH(\mathbf{x})\right| + \left|\int \langle \mathbf{t}, \mathbf{x}\rangle^2 \, dH(\mathbf{x})\right| \\
&\quad + \epsilon \|\mathbf{t}\| \int \langle \mathbf{t}, \mathbf{x}\rangle^2 \, d(F+G) + 3\int_{\Delta'} \langle \mathbf{t}, \mathbf{x}\rangle^3 \, d[F+G] \\
&\leqslant \left|\int \langle \mathbf{t}, \mathbf{x}\rangle \, dH(x)\right| + \|\mathbf{t}\|^2 \|C - D\|_s \\
&\quad + \epsilon \|\mathbf{t}\|^3 \int \|\mathbf{x}^2\| \, d(F+G) + 3\|\mathbf{t}\|^2 \int_{\Delta'} \|\mathbf{x}^2\| \, d(F+G).
\end{aligned} \tag{4}$$

Let G_{nk} denote the normal distribution with mean 0 and covariance matrix S_{nk}. Let $\{Y_{nk}: 1 \leqslant k \leqslant n, n \geqslant 1\}$ be a family of independent random variables with the distribution of Y_{nk} being G_{nk}. In addition, take $\{Y_{nk}\}$ to be independent of $\{U_{nk}\}$. It is easy to see that $Y_n = Y_{n_1} + \ldots + Y_{n_n}$ is asymptotically normal with mean 0 and covariance matrix S.

Let f_{nk}, f_n, g_{nk} and g_n denote the characteristic function of U_{nk}, U_n, Y_{nk} and Y_n respectively. Let

$$f^*_{nk}(\mathbf{t}) = E(e^{i\langle \mathbf{t}, U_{nk}\rangle} | V_{nk}).$$

To prove the lemma it is clearly sufficient to prove that, for each fixed $\mathbf{t}$,

$$\lim_{n\to\infty} |f_n(\mathbf{t}) - g_n(\mathbf{t})| = 0.$$

Let

$$W_{nk} = U_{n_1} + \ldots + U_{n,k-1} + Y_{n,k+1} + \ldots + Y_{nn} \quad \text{for} \quad 1 < k < n,$$

$$W_{n_1} = Y_{n_2} + \ldots + Y_{nn}, \quad W_{nn} = U_{n_1} + \ldots + U_{n,n-1}.$$

Then

$$\begin{aligned} |f_n(\mathbf{t}) - g_n(\mathbf{t})| &= |E(e^{i\langle \mathbf{t}, \mathbf{U}_n\rangle} - e^{i\langle \mathbf{t}, \mathbf{Y}_n\rangle}| \\ &= \left| E \sum_{k=1}^{n} (e^i\langle \mathbf{t}, \mathbf{U}_{nk}\rangle - e^i\langle \mathbf{t}, \mathbf{Y}_{nk}\rangle)\, e^i\langle \mathbf{t}, \mathbf{W}_{nk}\rangle \right| \\ &\leqslant \sum_{k=1}^{n} E|f^*_{nk}(\mathbf{t}) - g_{nk}(\mathbf{t})|. \end{aligned} \tag{5}$$

From (4) and (5), and the fact that $E(U_{nk}|V_{nk}) = 0$ we obtain

$$\begin{aligned} |f_n(\mathbf{t}) - g_n(\mathbf{t})| \leqslant \|t\|^2 \sum_{k=1}^{n} \|Er_{nk} - s_{nk}\|_s + 2\epsilon\|\mathbf{t}\|^3 \sum_{k=1}^{n} E\|\mathbf{U}_{nk}\|^2 \\ + 3\|\mathbf{t}\|^2 \sum_{k=1}^{n} E(\phi^\epsilon_{nk}\|\mathbf{U}_{nk}\|^2) + 3\|\mathbf{t}\|^2 \sum_{k=1}^{n} \int_{\Delta'} \|\mathbf{x}^2\|\, dG_{nk}. \end{aligned} \tag{6}$$

As $n \to \infty$ the first and third terms on the right of (6) go to 0 because of (1) and (3), the second term is $O(\epsilon)$ because of (2), and the last term goes to zero because G_{nk} is normal with covariance matrix S_{nk} and S_{nk} goes to 0 as $n \to \infty$ uniformly in $k \leqslant n$. Since ϵ is arbitrary, this completes the proof of the theorem.

Problem 1. Let $\{U_{nk};\ 1 \leqslant k \leqslant n, n \geqslant 1\}$ be a family of random variables defined on a probability space $(\Omega, \mathscr{A}, P)$ and taking values in the p-dimensional Euclidean space R^p. Let

$$(\beta_{nk};\ 1 \leqslant k \leqslant n,\ n \geqslant 1)$$

be a family of sub-G-fields of $\mathscr{A}$ such that, for each n, $\beta_{n_1} = (\Omega, \phi)$ while for $2 \leqslant k \leqslant n$, $\{(U_{n_1}, \ldots, U_{nk-1})\}$ is measurable with respect to β_{nk}. Suppose that

$$E(U_{nk}/\beta_{nk}) = 0 \quad \text{(almost everywhere)},$$

$$\lim_{n\to\infty} \Sigma E\|E[U_{nk}U'_{nk}/\beta_{nk}] - E(U_{nk}U'_{nk})\|_s = 0,$$

$$\sup_{1\leqslant n\leqslant\infty} \Sigma E\|U_{nk}\|_s^2 < \infty,$$

$$\lim_{n\to\infty} \Sigma \int_{[\|U_{nk}\|_s > \epsilon]} \|U_{nk}\|^2\, dp = 0 \quad \text{for every} \quad \epsilon > 0$$

and $$\lim_{n\to\infty} \Sigma E(U_{nk} U'_{nk}) = \mathbf{\Sigma},$$

where all the summations are over k such that $1 \leqslant k \leqslant n$. Then we have

(i) $\Sigma U_{nk} \to N(0, \mathbf{\Sigma})$;

(ii) $\Sigma U_{nk} U'_{nk} \to \mathbf{\Sigma}$ in probability, as $n \to \infty$.

4. Some theorems on conditional expectation

Kolmogorov inequality.

If $U_1, U_2, \ldots, U_n$ are random variables such that

$$E[U_i/U_1, U_2, \ldots, U_{i-1}] = U_{i-1} \quad (i = 2, 3, \ldots, n),$$

then for each $\epsilon > 0$

$$pr\{\sup_{1\leqslant i\leqslant n} \{|U_i| \geqslant \epsilon\} \leqslant \frac{EU_n^2}{\epsilon^2}.$$

Proof. Let $A_i = [|U_i| \geqslant \epsilon]$. Then

$$[\sup_{1\leqslant i\leqslant n} |U_i| \geqslant \epsilon] = \bigcup_{i=1}^{n} A_i = \bigcup_{i=1}^{n} B_i,$$

where $B_1 = A_1$, $B_i = A_i A_1^c A_2^c, \ldots, A_{i-1}^c$ $(i > 1)$. We define a pairwise disjoint sets $B_1, B_2, \ldots, B_n$ for $i = 1, 2, \ldots, n$

$$E[U_n^2|B_i] = E[\{U_i - (U_i - U_n)\}^2|B_i] = E[U_i^2|B_i]$$
$$+ E[(U_i - U_n)^2|B_i] + 2E[U_i(U_n - U_i)|B_i].$$

Note that

$$E[U_i^2|B_i] \geqslant \epsilon^2 \quad \text{since} \quad B_i \subset A_i \quad (i = 1, 2, \ldots, n),$$
$$E[(U_i - U_n)^2|B_i] \geqslant 0 \quad (i = 1, 2, \ldots, n).$$

If $i = n$, then $2E[U_i(U_n - U_i)|B_i] = 0$.

Suppose i $< n$

$$E[U_i(U_n - U_i)|B_i] = E\{E[U_i(U_n - U_i)|B_i, U_1, \ldots, U_{n-1}]\}$$
$$= E\{U_i(E[U_n|B_i, U_1, \ldots, U_{n-1}] - U_i)|B_i\}$$
$$= E[U_i(U_{n-1} - U_i)|B_i] = 0; \text{ if } i = n-1.$$

Proceeding in this way we obtain

$$E(U_n^2|B_i) \geqslant \epsilon^2 \quad \text{for} \quad i = 1, 2, \ldots, n,$$

then $$E(U_n^2) = \sum_{i=1}^{n} E[U_n^2/B_i]p(B_i) \geqslant \epsilon^2 \sum_{i=1}^{n} p(B_i)$$
$$= \epsilon^2 p\left(\bigcup_{i=1}^{n} B_i\right).$$

Since B_i's are disjoint

$$= \epsilon^2 p[\sup_{1\leqslant i\leqslant n} |U_i| \geqslant \epsilon] \leqslant E(U_n^2)$$

or $$p[\sup_{1\leqslant i\leqslant n} |U_i| \geqslant \epsilon] \leqslant \frac{E(U_n^2)}{\epsilon^2}.$$

Corollary. *Suppose $X_1, X_2, \ldots, X_n$ are random variables such that $E[X_j|X_1, \ldots, X_{j-1}] = 0$ for $j = 2, 3, \ldots, n$, then, for each $\epsilon > 0$*

$$p\left[\sup_{1\leqslant i\leqslant n}\left|\sum_{j=1}^{i} X_j\right| \geqslant \epsilon\right] \leqslant \frac{E\left[\sum_{j=1}^{n} X_j\right]^2}{\epsilon^2}.$$

Proof. Let $$U_i = \sum_{j=1}^{i} X_j$$

$$E[U_i|U_1, \ldots, U_{i-1}] = U_{i-1}.$$

First we note that $X_1, \ldots, X_{i-1}$ 'determine' $U_1, \ldots, U_{i-1}$ and in turn $X_1, \ldots, X_{i-1}$ are determined by $U_1, \ldots, U_{i-1}$, since

$$X_1 = U_1, \quad X_2 = U_2 - U_1, \quad X_i = U_{i-1} - U_{i-2}.$$

Thus

$$E[U_i|U_1, \ldots, U_{i-1}] = E[U_i|X_1, \ldots, X_{i-1}]$$
$$= E\left[\sum_{j=1}^{i-1} X_j + X_i|X_1, \ldots, X_{i-1}\right]$$
$$= E\left[\sum_{j=1}^{i-1} X_j|X_1, \ldots, X_{i-1}\right] + E[X_i|X_1, \ldots, X_{i-1}]$$
$$= \sum_{j=1}^{i-1} X_j + 0 = U_{i-1}.$$

Therefore previous inequality can be applied.

Special case.

If $X_1, \ldots, X_n$ are independent and $EX_j = 0$ $j = 1, 2, \ldots, n$ then

$$p\left[\sup_{1 \leqslant i \leqslant n} \left|\sum_{j=1}^{i} X_j\right| \geqslant \epsilon\right] \leqslant \sum_{j=1}^{n} \operatorname{var} X_j/\epsilon^2.$$

Tchebichev's inequality implies

$$p\left[\left|\sum_{j=1}^{n} X_j\right| \geqslant \epsilon\right] \leqslant \sum_{j=1}^{n} \operatorname{var} X_j/\epsilon^2.$$

LEMMA 1. *If X_n is a random variable sequence such that*

$$\sum_{n=1}^{\infty} EX_n^2 < \infty,$$

then the random variable sequence

$$\sum_{i=1}^{n} [X_i - E(X_i|X_1, \ldots, X_{i-1})] \quad n = 1, 2, \ldots$$

converges to a random variable with probability one.

Proof. Let $\epsilon > 0$ and $\eta > 0$. Let m_0 be a positive number such that

$$\sum_{n=m_0}^{\infty} EX_n^2 < \eta\epsilon^2.$$

Let m and n satisfy $m_0 \leqslant m < n$. We shall show that

$$p\left\{\sup_{m \leqslant j \leqslant n} \left|\sum_{i=m+1}^{j} [X_i - E(X_i|X_i, \ldots, X_{i-1})]\right| \geqslant \epsilon\right\} < \eta,$$

this implies the result.

Consider the random variables

$$Y_1 = X_{m+1} - E[X_{m+1}|X_1, \ldots, X_m],$$

$$Y_2 = X_{m+2} - E[X_{m+2}|X_1, \ldots, X_{m+1}],$$

$$Y_{n-m} = X_n - E[X_n|X_1, \ldots, X_{n-1}].$$

To prove that $E[Y_j|Y_1, \ldots, Y_{j-1}] = 0$ $j = 2, \ldots, n-m$. We shall prove that $*E[Y_j|Y_1, \ldots, Y_{j-1}, X_1, \ldots, X_m] = 0$ which will imply

$$E[Y_j|Y_1, \ldots, Y_{j-1}] = E_{X's}[E(Y_j|Y_1, \ldots, Y_{j-1}, X_1, \ldots, X_m)]$$

$$= E_{X's}(O) = 0.$$

If $j > 1$, then $X_1, X_2, \ldots, X_m, X_{m+1}, \ldots, X_{m+j-1}$ 'determine'

$$Y_1, Y_2, \ldots, Y_{j-1}, X_1, \ldots, X_m$$

and vice versa. Hence

$$\begin{aligned} &E[Y_j|Y_1, \ldots, Y_{j-1}, X_1, \ldots, X_n] \\ &\quad = E[Y_j|X_1, \ldots, X_m, X_{m+1}, \ldots, X_{m+j-1}] \\ &\quad = E\{X_{m+j} - E[X_{m+j}|X_1, \ldots, X_{m+j-1}]|X_1, \ldots, X_{m+j-1}\}, \\ &* = E[X_{m+j}|X_1, \ldots, X_{m+j-1}] - E[X_{m+j}|X_1, \ldots, X_{m+j-1}] = 0 \end{aligned}$$

and * is proved.

Thus we can apply the corollary to Kolmogorov inequality and get that

$$\begin{aligned} &p\left\{\sup_{m \leqslant j \leqslant n}\left|\sum_{i=m+1}^{j}(X_i - E(X_i|X_i, \ldots, X_{i-1}))\right| \geqslant \epsilon\right\} \\ &\quad \leqslant \frac{1}{\epsilon^2}E\left[\sum_{i=m+1}^{n}(X_i - E(X_i|X_1, \ldots, X_{i-1}))\right]^2 \\ &\quad = \frac{1}{\epsilon^2}\left[\sum_{i=m+1}^{n}E(X_j - E(X_i|X_1, \ldots, X_{i-1}))^2\right] \\ &\qquad + 2\sum_{m \leqslant i < j \leqslant n}E[(X_i - E(X_i|X_1, \ldots, X_{i-1})) \\ &\qquad\qquad \times (X_j - E(X_j|X_1, \ldots, X_{j-1}))]. \end{aligned}$$

Now if $i < j$

$$\begin{aligned} &E\{(X_i - E(X_i|X_1, \ldots, S_{i-1}))(X_j - E(X_j|X_1, \ldots, \ldots X_{j-1}))\} \\ &\quad = E\{E[(X_i - E(X_i|X_1, \ldots, X_{i-1}))(X_j - E(X_j|X_1, \ldots, X_{j-1}))/ \\ &\qquad\qquad X_1, \ldots, X_{j-1}]\} = 0. \end{aligned}$$

Also $\quad E(X_i - E(X_i|X_1, \ldots, X_{i-1}))^2$

$$\begin{aligned} &= E\{E[(X_i - E(X_i|X_1, \ldots, X_{i-1}))]^2|X_1, \ldots, X_{i-1}\} \\ &\leqslant E\{E[X_i^2|X_1, \ldots, X_{i-1}]\} \\ &= E(X_i^2), \end{aligned}$$

$$\begin{aligned} &p\left\{\sup_{m \leqslant j \leqslant n}\left|\sum_{i=m+1}^{n}(X_i - E(X_i X_1, \ldots, X_{i-1}))\right| \geqslant \epsilon\right\} \\ &\qquad \geqslant \frac{1}{\epsilon^2}\sum_{i=m+1}^{n}E(X_i^2) \leqslant \frac{1}{\epsilon^2}\sum_{i=m_0}^{\infty}E(X_i^2) \leqslant \frac{1}{\epsilon^2}\eta\epsilon^2 = \eta. \end{aligned}$$

This completes the proof.

LEMMA. *Suppose* $\{s_i\}$ *is a convergent real number sequence and that* $\{\alpha_i\}$ *is a positive number sequence such that* $\sum_i \alpha_i$ *diverges then*

$$\lim_{n\to\infty} \frac{\sum_{i=1}^{n} \alpha_i s_i}{\sum_i^n \alpha_i} = \lim_{n\to\infty} s_n.$$

Proof. Let $S = \lim_{n\to\infty} s_n$. Let $M = \sup_i |s_i - S|$ (finite). Let $\epsilon > 0$. Let m_1 be a positive integer such that if $i > m_1$ then $|s_i - S| < \epsilon/2$. Let $\epsilon/2$ be a positive integer $> m_1$ and such that

$$M \frac{\sum_{i=1}^{m_1} \alpha_i}{\sum_{i=1}^{n} \alpha_i} < \epsilon/2 \quad \text{if} \quad n > m_2.$$

Then if $n > m_2$

$$\left| \frac{\sum_{i=1}^{n} \alpha_i s_i}{\sum_{i=1}^{n} \alpha_i} - S \right| = \left| \frac{\sum_{i=1}^{n} \alpha_i (s_i - S)}{\sum_1^n \alpha_i} \right|$$

$$\leqslant \frac{\Sigma \alpha_i |(s_i - S)|}{\sum_{i=1}^{n} \alpha_i}$$

$$\leqslant M \frac{\sum_{i=1}^{m} \alpha_i}{\sum_1^n \alpha_i} + \epsilon/2 \frac{\sum_{i=m+1}^{n} \alpha_i}{\sum_{i=1}^{n} \alpha_i} < \epsilon/2 + \epsilon/2 = \epsilon.$$

KRONECKER'S LEMMA. *If* $0 < a_n \uparrow \infty$ *and* $\{x_n\}$ *is a real number sequence such that* $\sum_n (x_n/a_n)$ *converges, then*

$$\frac{1}{a_n} \sum_{i=1}^{n} x_i \to 0 \quad \text{as} \quad n \to \infty.$$

Proof. Let $s_0 = 1$ and if $n \geqslant 1$ let

$$s_n = \sum_{i=1}^{n} \frac{x_i}{a_i}.$$

Since $\lim_{n\to\infty} s_n$ exists, we have that

$$\lim_{n\to\infty} \frac{a_1 s_0 + \sum_{i=1}^{n-1} (a_{i+1} - a_i) s_i}{a_n} = \lim_{n\to\infty} s_n$$

or since $a_n \to \infty$

$$\lim_{n\to\infty} \left[\frac{\sum_{i=1}^{n-1} (a_{i+1} - a_i) s_i}{a_n} - s_n \right]$$

$$= \lim_{n\to\infty} \left[\frac{\sum_{i=2}^{n} \left(a_i s_{i-1} + \sum_{i=2}^{n} a_i \frac{x_i}{a_i} - \sum_{1}^{n} x_i - \sum_{i=1}^{n} a_i s_i + a_n s_n + x_1 - a_1 s_1 \right)}{a_n} - s_n' \right]$$

$$= \lim_{n\to\infty} \frac{1}{a_n} \sum_{1}^{n} x_i = 0 \quad \text{follows from preceding lemma.}$$

LEMMA 2. *If $\{X_n\}$ is a random variable sequence such that*

$$\sum_{n=1}^{\infty} \frac{X_n^2}{n^2} < \infty$$

then

$$\frac{1}{n} \sum_{i=1}^{n} [X_i - E(X_i | X_1, \ldots, X_{i-1})] \to 0 \quad \textit{with probability one.}$$

Proof. The result follows from Kronecker's lemma and lemma 1.

THEOREM 7. *If $\lim_{n\to\infty} (X_n) = X$ with probability one, and if there is a random variable $U \geqslant 0$, with $E(U) < \infty$, such that $|X_n| < U$ with probability one, then*

$$\lim_{n\to\infty} [E\{X_n | F\}] = E(X|F),$$

where F is Borel field.

Proof. To prove the theorem define $\hat{X}_n$ by

$$\hat{X}_n = \underset{j \geqslant n}{\text{L.U.B.}} |X_j - X|$$

then $$\hat{X}_1 > \hat{X}_2 > \hat{X}_3 > \ldots \geqslant 0, \hat{X}_n \leqslant 2U$$

with probability one and

$$\lim_{n\to\infty} \hat{X}_n = 0 \quad \text{with probability one} \quad (\text{given } \lim_{n\to\infty} X_n = X$$

$$\text{with probability one})$$

$$|E(X|F) - E(X_n|F)| = |E\{(X - X_n)|F\}| \leqslant E\{|X - X_n|\,|F\}$$
$$\leqslant E\{\hat{X}_n|F\} \quad \text{with probability one.}$$

Hence it is sufficient to prove that

$$\lim_{n\to\infty} E\{\hat{X}_n|F\} = 0 \quad \text{with probability one.}$$

$$E(\hat{X}_1|F) \geqslant E(\hat{X}_2|F) \ldots \geqslant 0 \quad \text{with probability one,}$$

so that there must be convergence.

If $\lim_{n\to\infty} E(X_n/F) = W$ with probability one

$$E\{W\} \leqslant E\{E(\hat{X}_n|F)\} = E(\hat{X}_n)$$

and the right side goes to 0 when $n \to \infty$ because it is the integral of $\hat{X}_n$, where $\hat{X}_n$ is dominated by $2U$ and goes to 0 with probability one when $n \to \infty$. Thus $E(W) = 0$, so that $W = 0$ with probability one, as was to be proved.

Let F_1 and F_2 be Borel fields of measurable W sets with $F_1 \subset F_2$ then

$$E[E(X|F_2)|F_1] = E(X|F_1)$$

with probability one. For example, if $X, X_1, X_2, \ldots,$ are random variables with $E\{|X|\} < \infty$ then

$$E[E\{X|X_1, X_2, \ldots,\}|\{X_2, X_4, \ldots\}] = E(X|X_2, X_4, \ldots)$$

with probability one.

5. Problems

1. Let $\{X_n\}$ be a sequence of random variables satisfying

   (i) $\sup_n E\{|X_n|\} < \infty$;

   (ii) $\sum_{n=1}^{\infty} E\{[E\{X_{n+1} - X_n|X_1, \ldots, X_n\}^+]\} \to \infty,$

then prove X_n converges almost surely to random variable as $n \to \infty$, where $X = \frac{1}{2}[X+|X|]$.

Hint. Use Martingale convergence theorem of Doob [22]. (Blum)

2. Let $\{X_n\}$ be a sequence of integrable random variables which satisfy conditions (ii) of the previous problem and are bounded below uniformly in n. Then X_n converges almost surely to a random variable where $n \to \infty$.

Hint. Let $Y_n = X_n - a$, where a is chosen so that $Y_n \geqslant 0$ for all n. (Blum)

3. Let $\{X_n\}$ be a sequence of real valued random variables and $\{\beta_n\}$ be a sequence of σ-fields such that $X_1, \ldots, X_{n-1}$ is measureable with respect to β_n for $n > 1$. Prove

(i) If $\sum_n EX_n^2 < \infty$ and $\sum_n E[X_n|\beta_n]$ converges a.s., then $\sum_n X_n$ converges a.s.;

(ii) If $\sum_n b_n^{-2} E(X_n^2) < \infty$ with $b_n \uparrow \infty$ then

$$b_n^{-1} \sum_{k=1}^{n} \{X_k - E[X_k|\beta_k]\} \to 0 \text{ a.s.} \quad \text{as} \quad n \to \infty. \qquad \text{(Venter)}$$

APPENDIX 3

INEQUALITIES

1. Introduction

In this appendix we shall prove some inequalities which are used in solving various aspects of the problems of stochastic-approximation processes. Now we give the following definitions.

DEFINITION 1.

$$\alpha_n = O(\beta_n) \Leftrightarrow \frac{\alpha_n}{\beta_n} \quad \textit{is a bounded sequence.}$$

DEFINITION 2.

$$\alpha_n = o(\beta_n) \Leftrightarrow \frac{\alpha_n}{\beta_n} \to 0 \quad as \quad n \to \infty.$$

DEFINITION 3.

$$\alpha_n = o(1) \Leftrightarrow \alpha_n \to 0 \quad as \quad n \to \infty.$$

2. Inequalities developed for stochastic approximation

LEMMA 1 (Chung). *Suppose that* $\{\xi_n\}$ *is a real number sequence such that for* $n \geqslant n_0$,

$$\xi_{n+1} \leqslant \left(1-\frac{c}{n}\right)\xi_n + \frac{d}{n^{p+1}},$$

where $c > p > 0$, $d > 0$ *then*

$$\xi_n \leqslant \frac{d}{c-p}\frac{1}{n^p} + O\left(\frac{1}{n^p}\right).$$

Proof. We have that for $|X| < 1$

$$(1+X)^{-p} = 1 + \binom{-p}{1}X + \binom{-p}{2}X^2 \ldots = 1 - pX + o(X^2)$$

as $X \to 0$.

Therefore $n > 1$

$$\frac{1}{n^p}-\frac{1}{(n+1)^p}=\frac{1}{n^p}-\frac{1}{n^p}\left[1+\frac{1}{n}\right]^{-p}$$
$$=\frac{1}{n^p}-\frac{1}{n^p}\left[1-\frac{p}{n}+o\left(\frac{1}{n^2}\right)\right]$$
$$=\frac{p}{n^{p+1}}+o\left(\frac{1}{n^{p+2}}\right),$$

where $o(1/n^{p+2})$ is positive therefore

$$\frac{1}{(n+1)^p}-\left(1-\frac{c}{n}\right)\frac{1}{n^p}=\frac{c}{n^{p+1}}-\left(\frac{1}{n^p}-\frac{1}{(n+1)^p}\right)$$
$$=\frac{c-p}{n^{p+1}}+o\left(\frac{1}{n^{p+2}}\right),$$

here 0 is negative;

$$\frac{d}{c-p}\left[\frac{1}{(n+1)^p}-\left(1-\frac{c}{n}\right)\frac{1}{n^p}\right]=\frac{d}{n^{p+1}}+o\left(\frac{1}{n^{p+2}}\right)$$
$$\geqslant\frac{d}{n^{p+1}}-\frac{k_1}{n^{p+2}},$$

where $$k_1=\sup_n\left[\frac{o(1/n^{p+2})}{1/n^{p+2}}\right].$$

Let $0<\epsilon<c-p$. Then $c-p-\epsilon>0$ and $0<\epsilon\leqslant 1$ also

$$\frac{1}{(n+1)^{p+\epsilon}}-\left(1-\frac{c}{n}\right)\frac{1}{n^{p+\epsilon}}=\frac{c-p-\epsilon}{n^{p+\epsilon+1}}+o\left(\frac{1}{n^{p+\epsilon+2}}\right).$$

There is a $k_2>0$ such that for $n\geqslant n_1$

$$\frac{k_1}{n^{p+2}}\leqslant k_2\left[\frac{1}{(n+1)^{p+\epsilon}}-\left(1-\frac{c}{n}\right)\frac{1}{n^{p+\epsilon}}\right],$$

since $p+\epsilon+1<p+2$ and $c-p-\epsilon>0$. Thus for

$$n\geqslant\max\,[n_0,n_1]$$

$$\xi_{n+1}\leqslant\left(1-\frac{c}{n}\right)\xi_n+\frac{d}{n^{p+1}}$$
$$\leqslant\left(1-\frac{c}{n}\right)\xi_n+\frac{d}{c-p}\left[\frac{1}{(n+1)^p}-\left(1-\frac{c}{n}\right)\frac{1}{n^p}\right]+\frac{k_1}{n^{p+2}}$$

$$\leqslant \left(1-\frac{c}{n}\right)\xi_n + \frac{d}{c-p}\left[\frac{1}{(n+1)^p} - \left(1-\frac{c}{n}\right)\frac{1}{n^p}\right]$$

$$+k_2\left[\frac{1}{(n+1)^{p+\epsilon}} - \left(1-\frac{c}{n}\right)\frac{1}{n^{p+\epsilon}}\right],$$

$$\xi_{n+1} - \frac{d}{c-p}\frac{1}{n^{p+1}} - \frac{k_2}{(n+1)^{p+\epsilon}} \leqslant \left(1-\frac{c}{n}\right)\left[\xi_n - \frac{d}{c-p}\frac{1}{n^p} - \frac{k_2}{n^{p+\epsilon}}\right].$$

Denote left-hand side by ξ'_{n+1} and right-hand side by

$$\xi'_n\left(1-\frac{c}{n}\right), \quad \xi'_{n+1} \leqslant \xi'_n.$$

If for some $n > c, n > \max[n_0, n_1]\, \xi'_n < 0$, this implies the desired inequality is true for every n, i.e.

$$\xi_n \leqslant \frac{d}{c-p}\cdot\frac{1}{n^p} + \frac{k_2}{n^{p+\epsilon}}.$$

Otherwise for every $n > n_1 > \max[c, n_0, n_1]$ we have

$$0 < \xi'_n \leqslant \xi'_{n-1}\left[1-\frac{c}{n-1}\right]$$

$$\leqslant \xi'_{n_1}\prod_{m=n_1}^{n-1}\left(1-\frac{c}{m}\right)$$

$$\leqslant \xi'_{n_1}\prod_{m=n_1}^{n-1} e^{-c/m}$$

$$\leqslant \xi'_{n_1} e^{-c}\sum_{n_1}^{n-1}\frac{1}{m}$$

$$= \xi'_{n_1} e^{-c}\sum_{n_1}^{n-1}\frac{1}{m} \approx \xi'_{n_1} e^{-c\log n} = \xi'_{n_1}\frac{1}{n^c}.$$

Therefore for $n > n_1$

$$\xi_n \leqslant \frac{d}{c-p}\frac{1}{n^p} + \frac{k_2}{n^{p+\epsilon}} + \xi'_{n_1}\frac{1}{n^c}.$$

Thus in either case

$$\xi_n \leqslant \frac{d}{c-p}\frac{1}{n^p} + o\left(\frac{1}{n^p}\right).$$

LEMMA 2 (Burkholder). *Suppose* $\{\xi_n\}$ *is a non-negative number sequence and each of* c_n *and* d_n *is a real number sequence such that for all* $n \geqslant n_0$

$$\xi_{n+1} \leqslant \xi_n \left(1 - \frac{c_n}{n}\right) + \frac{d_n}{n^{p+1}},$$

where $\liminf_{n\to\infty} c_n = c > p > 0$ *and* $\limsup_{n\to\infty} d_n = d > 0$

then

$$\overline{\lim_{n\to\infty}}\, n^p \xi_n \leqslant \frac{d}{c-p}$$

or equivalently

$$\xi_n \leqslant \frac{d}{c-p} \cdot \frac{1}{n^p} + o\left(\frac{1}{n^p}\right).$$

Proof. Let $c - p > \epsilon > 0$ by assumption. There is an $n_1 > n_0$ such that if $n > n_1$ then $c_n > c - \epsilon > p$ and $d_n < d + \epsilon$. Thus, since $\{\xi_n\}$ is non-negative for $n > n_1$

$$\xi_{n+1} \leqslant \xi_n \left(1 - \frac{c_n}{n}\right) + \frac{d_n}{n^{p+1}}$$

$$\leqslant \xi_n \left(1 - \frac{c-\epsilon}{n}\right) + \frac{d+\epsilon}{n^{p+1}}$$

which implies by the lemma 1, since $c - \epsilon > p > 0$ and $d + \epsilon > 0$

$$\lim_{n\to\infty} n^p \xi_n \leqslant \frac{d+\epsilon}{c-\epsilon-p}.$$

Since ϵ is any positive number, as $\epsilon \to 0$

$$\limsup_{n\to\infty} n^p \xi_n \leqslant \frac{d}{c-p}.$$

LEMMA 3 (Chung). *Suppose that* $\{\xi_n\}$ *is a real number sequence such that for* $n \geqslant n_0$

$$\xi_{n+1} \geqslant \left(1 - \frac{c}{n}\right) \xi_n + \frac{d}{n^{p+1}},$$

where $c > p > 0, d > 0$ *then*

$$\xi_n \geqslant \frac{d}{c-p} \cdot \frac{1}{n^p} + o\left(\frac{1}{n^p}\right).$$

Proof. As before,

$$\frac{1}{(n+1)^p} - \left(1 - \frac{c}{n}\right) \frac{1}{n^p} = \frac{c-p}{n^{p+1}} + o\left(\frac{1}{n^{p+2}}\right).$$

For all real number c and p, and thus for $c > p > 0$, $d > 0$

$$\frac{d}{c-p}\left[\frac{1}{(n+1)^p} - \left(1-\frac{c}{n}\right)\frac{1}{n^p}\right] = \frac{d}{n^{p+1}} + \frac{d}{c-p}O\left(\frac{1}{n^{p+2}}\right) \leqslant \frac{d}{n^{p+1}}$$

since $O(1/n^{p+2})$ is negative. Thus for $n \geqslant n_0$

$$\begin{aligned}\xi_{n+1} &\geqslant \xi_n\left(1-\frac{c}{n}\right) + \frac{d}{n^{p+1}} \\ &\geqslant \xi_n\left(1-\frac{c}{n}\right) + \frac{d}{c-p}\left[\frac{1}{(n+1)^p} - \left(1-\frac{c}{n}\right)\frac{1}{n^p}\right],\end{aligned}$$

$$\xi_{n+1} - \frac{d}{c-p}\frac{1}{(n+1)^p} \geqslant \left(1-\frac{c}{n}\right)\left[\xi_n - \frac{d}{c-p}\frac{1}{n^p}\right],$$

$$\xi'_{n+1} \geqslant \left(1-\frac{c}{n}\right)\xi'_n.$$

If for some $n > c$, $n > n_0$, $\xi'_n \geqslant 0$ then $\xi'_{n+1} > 0$ for all subsequent n, namely

$$\xi_n \geqslant \frac{d}{c-p}\frac{1}{n^p},$$

otherwise for every $n > \max[c, n_0]$

$$\begin{aligned}0 > \xi'_n > \xi'_{n+1}\left(1-\frac{c}{n-1}\right) &\geqslant \xi'_{n-2}\left(1-\frac{c}{n-2}\right)\left(1-\frac{c}{n-1}\right) \\ &\geqslant \xi'_{n_1}\prod_{m=n_1}^{n-1}\left(1-\frac{c}{m}\right) \\ &\approx \xi'_{n_1}\frac{1}{n^c}.\end{aligned}$$

Thus for $n > c$, $n \geqslant n_0$

$$\begin{aligned}\xi_n &\geqslant \frac{d}{c-p}\frac{1}{n^p} + \xi'_{n_1}\frac{1}{n^c} \\ &\geqslant \frac{d}{c-p}\frac{1}{n^p} + o\left(\frac{1}{n^p}\right).\end{aligned}$$

LEMMA 4 (Burkholder). *Suppose $\{\xi_n\}$ is a non-negative number sequence and each of $\{c_n\}$ and $\{d_n\}$ is a real number sequence such that for all $n > n_0$*

$$\xi_{n+1} = \xi_n\left(1-\frac{c_n}{n}\right) + \frac{d_n}{n^{p+1}},$$

where $\lim_{n\to\infty} c_n = c > p > 0$ *and* $\lim_{u\to\infty} d_n = d > 0$

then
$$\lim_{n\to\infty} n^p \xi_n = \frac{d}{c-p}.$$

Proof. We have by lemma 2 that
$$\limsup_{n\to\infty} n^p \xi_n \leqslant \frac{d}{c-p}.$$
We shall now prove that
$$\liminf_{n\to\infty} n^p \xi_n \geqslant \frac{d}{c-p}$$
which will imply
$$\frac{d}{c-p} \geqslant \liminf_{n\to\infty} n^p \xi_n \leqslant \limsup_{n\to\infty} n^p \xi_n \leqslant \frac{d}{c-p}.$$
In other words desired result holds.

Let ϵ satisfy $d > \epsilon > 0$ by assumption there is an $n_1 > n_0$ such that if $n > n_1, c_n < c+\epsilon$ and $d_n > d-\epsilon$. Thus if $n > n_1$
$$\xi_{n+1} = \xi_n\left(1-\frac{c_n}{n}\right)+\frac{d_n}{n^{p+1}} \geqslant \xi_n\left(1-\frac{c+\epsilon}{n}\right)+\frac{d-\epsilon}{n^{p+1}}.$$
By lemma 3
$$\liminf_{n\to\infty} n^p \xi_n \geqslant \frac{d-\epsilon}{c+\epsilon-p} \to \frac{d}{c-p} \quad \text{as} \quad \epsilon \to 0.$$

Thus the desired result.

Lemma 5 (Chung). *Suppose that* $\{\xi_n\}$, $n \geqslant 1$ *is a sequence of real numbers such that for* $n \geqslant n_0$
$$\xi_{n+1} \geqslant \left(1-\frac{c}{n^s}\right)\xi_n + \frac{c_1}{n^t}, \tag{1}$$
where $o < s < 1$, $s < t$, $c > 0$, $c_1 > 0$. *Then*
$$\liminf_{n\to\infty} n^{t-s} \xi_n \leqslant \frac{c_1}{c}.$$

Proof. We may take $n_0 = 1$. We have
$$\frac{1}{(n+1)^{t-s}} - \left(1-\frac{c}{n^s}\right)\frac{1}{n^{t-s}} \leqslant \frac{c}{n^t}.$$

Hence we have

$$\frac{c_1}{n^t} \geqslant \frac{c_1}{c}\left[\frac{1}{(n+1)^{t-s}} - \left(1-\frac{c}{n^s}\right)\frac{1}{n^{t-s}}\right].$$

Using (1) we obtain

$$\xi_{n+1} - \frac{c}{c(n+1)^{t-s}} \geqslant \left(1-\frac{c}{n^s}\right)\left(\xi_n - \frac{c_1}{cn^{t-s}}\right).$$

If for some $n > c^{1/s}$ we have $\xi_n \geqslant (c_1/cn^{t-s})$, then this is true for all subsequent n. Otherwise for every $n > n_1 > c^{1/s}$ we have

$$\left|\xi_n - \frac{c_1}{cn^{t-s}}\right| \leqslant \left|\xi_{n_1} - \frac{c_1}{cn_1^{t-s}}\right| \prod_{m=n_1}^{n-1}\left(1-\frac{c}{m^s}\right) = O\left(\frac{1}{n^q}\right)$$

for every $q > 0$. The lemma follows in either case.

LEMMA 6 (Chung). *Suppose that* $\{\xi_n\}$, $n \geqslant 1$ *is a sequence of real numbers such that for* $n \geqslant n_0$

$$\xi_{n+1} \leqslant \left(1-\frac{c_n}{n^s}\right)\xi_n + \frac{c'}{n^t}, \tag{1}$$

where $0 < s < 1$, $s < t$, $c_n \geqslant c > 0$, $c' > 0$. *Then*

$$\limsup_{n\to\infty} n^{t-s}\xi_n \leqslant \frac{c'}{c}.$$

Proof. We have if c'' is any number $> c'$

$$\begin{aligned}\frac{c'}{n^t} &\leqslant \frac{c''}{c_n}\left[\frac{1}{(n+1)^{t-s}} - \left(1-\frac{c_n}{n^s}\right)\frac{1}{n^{t-s}}\right]\\ &\leqslant \frac{c''}{c}\left[\frac{1}{(n+1)^{t-s}} - \left(1-\frac{c_n}{n^s}\right)\frac{1}{n^{t-s}}\right]\end{aligned}$$

for all $n > n_0(c'')$. Using (1), we obtain

$$\xi_{n+1} - \frac{c''}{c}\,\frac{1}{(n+1)^{t-s}} \leqslant \left(1-\frac{c_n}{n^s}\right)\left(\xi_n - \frac{c''}{c}\,\frac{1}{n^{t-s}}\right).$$

The rest follows as in the proof of lemma 5.

REMARK. If we replace c' by sequence $\{c'_n\}$ of positive real numbers such that $c'_n \to \infty$ as $n \to \infty$ then $\limsup_{n\to\infty} n^{t-s}\xi_n \leqslant 0$.

LEMMA 7 (Burkholder). *Suppose* $\{\xi_n\}$ *is a non-negative number sequence and each of* $\{c_n\}$ *and* $\{d_n\}$ *is a real number sequence suchthat*

$$\text{(i)}\quad \xi_{n+1} \leqslant \xi_n\left[1-\frac{c_n+o(1)}{n}\right]+\frac{d_n+o(1)}{n^{p+1}}+\frac{O(1)}{n^{q+1}};$$

$$\text{(ii)}\quad \xi_{n+1} \leqslant \xi_n\left[1-\frac{c_n}{n}\right]+\frac{d_n}{n^{p+1}}+\frac{O(1)\,\xi_n^r}{n^{q+1}};$$

where $\liminf_{n\to\infty} c_n = c > p > 0$, $\limsup_{n\to\infty} d_n = d \geqslant 0$, $0 < r < 1$ *and*

$$p(1-r) < q$$

then
$$\limsup_{n\to\infty} n^p\,\xi_n \leqslant \frac{d}{c-p}.$$

Proof. If $q > p$ then

$$\frac{d_n+o(1)}{n^{p+1}}+\frac{O(1)}{n^{q+1}} = \frac{d_n+o(1)}{n^{p+1}}$$

and the desired conclusion would be implied by the lemma 2 and (i). Hence, lemma 7 needs only to be proved for

$$p(1-r) < q \leqslant p; \quad q \leqslant p < q\left(\frac{1}{1-r}\right).$$

For this case lemma 2 implies, using (i), that

$$\overline{\lim_{n\to\infty}}\, n^q\,\xi_n \leqslant \frac{k}{c-q}$$

for some $k > 0$ or $\xi_n = O(n^{-q})$.

Thus
$$\xi_n^r = O(n^{-qr}) = \frac{O(1)}{n^{qr}}.$$

Substituting in (2) gives

$$\xi_{n+1} \leqslant \xi_n\left[1-\frac{c_n}{n}\right]+\frac{d_n}{n^{p+1}}+\frac{O(1)}{n^{q+qr+1}}.$$

If $q+qr > p$, then the desired result is implied, if

$$q(1+r) \leqslant p < \frac{q}{1-r}$$

let $$\xi_n = O\left(\frac{1}{n^{q(1+r)}}\right),$$

$$\xi_n^r = O\left(\frac{1}{n^{q(r+r^2)}}\right) = \frac{O(1)}{n^{q(r+r^2)}}.$$

Substituting in (2) gives

$$\xi_{n+1} \leqslant \xi_n\left[1-\frac{c_n}{n}\right]+\frac{d_n}{n^{p+1}}+\frac{O(1)}{n^{q(1+r+r^2)+1}}.$$

If $q(1+r+r^2) > p$ then the desired result is implied, if

$$q(1+r+r^2) \leqslant p < \frac{q}{1-r}$$

then continue. Eventually, since $p < q/(1-r)$, one will get to a k such that $q(1+r+r^2+\ldots+r^k) > p$.

Hence the desired result.

LEMMA 8 (Derman and Sacks). *Let* $\{a_n\}$, $\{b_n\}$, $\{c_n\}$, $\{\delta_n\}$ *and* $\{\xi_n\}$ *be sequences of real numbers satisfying*

(i) $\{a_n\}$, $\{b_n\}$, $\{c_n\}$, $\{\xi_n\}$ *are non-negative;*

(ii) $\lim_{n\to\infty} a_n = 0$, $\Sigma b_n < \infty$, $\Sigma c_n = \infty$, $\Sigma \delta_n$ *converges, and for all n larger than some* N_0;

(iii) $\xi_{n+1} \leqslant \max[a_n, (1+b_n)\xi_n+\delta_n-c_n]$ *then* $\lim_{n\to\infty}\xi_n = 0$.

Proof. Let $n_1 > N_0$ and write

$$B_n = \prod_{i=1}^{n}(1+b_i). \tag{1}$$

Take $n > n_1$ and interate (iii) back to n_1. This yields

$$\xi_{n+1} \leqslant \max\left(\frac{B_n}{B_{n_1-1}}\xi_{n_1}+B_n\sum_{j=n_1}^{n}\frac{\delta_j-c_j}{B_j},\right.$$

$$\left.\max_{n_1\leqslant k\leqslant n}\left[\frac{B_n}{B_k}a_k+B_n\sum_{j=k+1}^{n}\frac{\delta_j-c_j}{B_j}\right]\right). \tag{2}$$

Since $$B_n = \prod_1^n(1+b_i) < \sum_1^n b_i;\quad \sum_1^\infty \delta_n < \infty,$$

$$\sum_1^\infty b_i < \alpha\sum_{j=1}^\infty\frac{\delta_j}{b_j} < \infty \quad\text{and}\quad \sum_{j=1}^\infty\frac{c_j}{B_j} = \infty.$$

Since $(B_n/B_{n_1-1})\xi_{n_1}$ is finite we see that the first term in the right-hand side of (2) must be negative for large enough n and can therefore be ignored.

Thus, since (i) and (ii) imply that B_n increases to β for n large enough

$$\xi_{n+1} \leqslant \max_{n_1 \leqslant k \leqslant n} \left(\frac{B_n}{B_k} a_k + B_n \sum_{j=k+1}^{n} \frac{\delta_j - c_j}{B_j} \right)$$

$$\leqslant B \left(\max_{k > n_1} a_k + \max_{n_1 \leqslant k \leqslant n} \left| \sum_{j=k+1}^{n} \delta_{j/B_j} \right| \right). \tag{3}$$

Since $\Sigma(\delta_j/B_j)$ converges and $a_k \to 0$ the right member of (3) can be made arbitrarily small by choosing n_1 large enough. This completes the proof of the lemma.

LEMMA 9 (Derman and Sacks). *Let* $\{a_n\}$, $\{c_n\}$, $\{\xi_n\}$ *be as in lemma* 8. *Suppose*

(i) $\{\delta_n\}$ *are positive* $\Sigma\delta_n < \infty$;

(ii) $\sum_{n=1}^{\infty} b_n$ *converges*, $\sum_{n=1}^{\infty} b_n^2 < \infty$.

Then $\lim_{n\to\infty} \xi_n = 0$.

Proof. Let n_1 be large enough so that $|b_n| < 1$ for $n > n_1$. Since Σb_n converges we have, for $n, k > n_1$

$$0 < \frac{B_n}{B_k} = \prod_{i=k+1}^{n} (1+b_i) \leqslant \exp\left[\sum_{i=k+1} b_i\right]$$

$$\leqslant \exp\left[\max_{n \geqslant n_1}, \max_{n_1 \leqslant k < n} \left| \sum_{i=k+1}^{n} b_i \right| \right] \leqslant A < \infty. \tag{4}$$

Also, because of (ii) and the fact that

$$\frac{B_n}{B_{n_1}} \prod_{i=n_1+1}^{n} (1-b_i) = \prod_{i=n_1+1}^{n} (1-b_i^2),$$

we have

$$\lim_{n\to\infty} \frac{B_n}{B_{n_1}} > 0. \tag{5}$$

With (4) and (5) established the proof goes through as in lemma 8.

Lemma 10 (Venter). *Let $\{b_n\}$, $\{c_n\}$, $\{d_n\}$ be real number sequence such that*

$$\sum_n b_n \quad \textit{converges and} \quad \sum_n b_n^2 < \infty, \tag{1}$$

$$c_n \geqslant 0 \quad \sum_n c_n = \infty, \tag{2}$$

$$d_n \geqslant 0 \quad \sum_n d_n < \infty. \tag{3}$$

(*a*) *If $\{\xi_n\}$ is a sequence of non-negative numbers such that, for for some integer n_0 and all $n > n_0$,*

$$\xi_{n+1} \leqslant \max\,[a, (1+b_n)\,\xi_n + d_n - c_n], \tag{4a}$$

where $a > 0$ then

$$\limsup_{n\to\infty} \xi_n \leqslant a. \tag{5}$$

(*b*) *If, instead of (4a), $\{\xi_n\}$ satisfies*

$$\xi_{n+1} \leqslant \max\,[a, (1+b_n)\,\xi_n + d_n]. \tag{4b}$$

Then we can still conclude that the sequence $\{\xi_n\}$ is bounded.

Proof. Consider part (*a*) first. Let

$$\left.\begin{aligned} B_{k,n} &= \prod_{j=k+1} (1+b_j) \quad \text{for} \quad k < n, \\ &= 1 \qquad\qquad\qquad \text{for} \quad k = n. \end{aligned}\right\} \tag{6}$$

Let $0 < b < 1$. By (1), for some integer $m \geqslant n_0$ and all $n > m$, we have

$$|b_n| \leqslant b \tag{7}$$

using the inequality $1+t < \exp(t)$ which holds for all t, we get from (6) and (7), for $n, k \geqslant m$

$$0 < B_{k,n} < \exp\left[\sum_{j=k+1}^{n} b_j\right],$$

$$B_{k,n} \leqslant \exp\left[\max_{n>m}\, \max_{m\leqslant k\leqslant n} \left|\sum_{j=k+1}^{n} b_j\right|\right] = A_m, \quad \text{say.} \tag{8}$$

Similarly, for $n, k \geqslant m$

$$0 < \sum_{j=k+1}^{n} (1-b_j) \leqslant A_m \tag{9}$$

and it follows from (1) that

$$A_m \to 1 \quad \text{as} \quad m \to \infty. \tag{10}$$

Using the inequality $1-t > \exp[-t(1-t)^{-1}]$ which holds for $0 < t < 1$, we get from (7), for $k, n \geqslant m$

$$\prod_{j=k+1}^{n} (1-b_j^2) \geqslant \exp\left[-(1-b^2)^{-1} \sum_{j=m}^{n} b_n^2\right] = A'_m > 0, \quad \text{say.} \tag{11}$$

Hence, for all $n, k > m$

$$B_{k,n} = \prod_{j=k+1}^{n} (1-b_j^2) \bigg/ \prod_{j=k+1}^{n} (1-b_j) \geqslant \frac{A'_m}{A_m} = A''_m > 0. \tag{12}$$

Now, let $n > m$ and iterate $(4a)$ back to m. We get

$$\xi_{n+1} \leqslant \max\left\{\left[B_{m-1,n}\xi_m + \sum_{j=m}^{n} B_{j,n}(d_j-c_j)\right]^*,\right.$$
$$\left.\max_{m \leqslant k \leqslant n}\left[B_{k,n}a + \sum_{j=k+1}^{n} B_{j,n}(d_j-c_j)\right]\right\}. \tag{13}$$

By (8) and (3) $\quad \sum_{j=m}^{n} B_{j,n} d_j \leqslant A_m \sum_{j=m}^{n} d_j < \infty;$

and by (12) and (2)

$$\sum_{j=m}^{n} B_{j,n} c_j \geqslant A''_m \sum_{j=m}^{n} c_j \to \infty \quad \text{as} \quad n \to \infty.$$

Since $B_{m-1,n}\xi_m \leqslant A_{m-1}\xi_m < \infty$, it follows that for all n large enough the first term in the argument of max {*} in (13) becomes negative and can be ignored. Hence, as $n \to \infty$

$$\xi_{n+1} \leqslant \max_{m \leqslant k \leqslant n}\left[B_{k,n}a + \sum_{j=k+1}^{n} B_{j,n}(d_j-c_j)\right]$$
$$\leqslant A_m\left[a + \sum_{j=m}^{n} d_j\right],$$

i.e. $$\limsup_{n\to\infty} \xi_n \leqslant A_m\left[a + \sum_{j=m}^{\infty} d_j\right]$$

and, letting $m \to \infty$, using (10) and (3), (5) follows. Next, consider part (b); i.e. suppose $(4b)$ is satisfied, then letting $c_j = 0$ in (13) and using the inequalities

$$B_{m+1,n}\xi_n \leqslant A_{m-1}\xi_m \quad \text{and} \quad \sum_{j=k+1}^{n} B_{j,n} d_j \leqslant A_m \sum_{j=m}^{\infty} d_j$$

we obtain

$$\xi_{n+1} \leqslant \max\left[A_{m-1}\xi_m + A_m \sum_{j=m}^{\infty} d_j, A_m\left(a + \sum_{j=m}^{\infty} d_j\right)\right]$$

for all $n > m$. Since the right-hand side here is finite and independent of n, it follows that $\{\xi_n\}$ is bounded.

LEMMA 11 (Sacks). *Let $\{\delta_k\}$ be a sequence of real numbers converging to δ where δ may be taken to be ∞. Let $\{a_n\}$ be sequence of positive real numbers such that $\sum_n a_n = \infty$, $\sum_n a_n^2 < \infty$ and*

$$\begin{aligned} B_{k,n} &= \prod_{j=k+1}^{n} (1-a_j) \quad \textit{for} \quad 0 \leqslant k < n, \\ &= 1 \quad\quad\quad\quad\quad\quad \textit{for} \quad k = n. \end{aligned}$$

Then, for any positive integer k_0

$$\lim_{n\to\infty} \sum_{k=k_0}^{n} a_k B_{k,n} \delta_k = \delta.$$

Proof. For any fixed k it follows from the problem of Appendix 3 that $\lim_{n\to\infty} B_{k,n} = 0$. Since $a_k B_{k,n} = B_{k,n} - B_{k-1,n}$ we have for any fixed k_1,

$$\lim_{n\to\infty} \sum_{k=k_1}^{n} a_k B_{k,n} = \lim_{n\to\infty} (1 - B_{k_1-1,n}) = 1 \tag{1}$$

and for every

$$\epsilon > 0 \,\exists\, k_2 > 0 \ni |\delta_k - \delta| < \epsilon \quad \text{for} \quad k > k_2. \tag{2}$$

Take $k_0 = \max(k_1, k_2)$. Then for (1) and (2) it follows that

$$\lim_{n\to\infty} \sum_{k=k_0}^{\infty} a_k B_{k,n} \delta_k = \delta.$$

Hence the desired lemma.

LEMMA 12 (Sacks). *Let $\{c_n\}$ be a sequence of positive real numbers. Let $a > \frac{1}{2}$ and $a_n = an^{-1}$ and for each n let*

$$h_n = \left(\sum_{k=1}^{n} a^2 C_k^{-2} k^{-2} \beta_{k,n}^2\right)^{-\frac{1}{2}}.$$

Suppose that $c_k \leqslant c < \infty$ for all k. Then, if k_0 is some fixed positive integer

$$\lim_{n\to\infty} h_n B_{k_0,n} = 0.$$

Proof. If $k_0 > a-1$ let $k_1 = k_0$; otherwise, let k_1 be the smallest integer greater than $a-1$. To prove the lemma it is sufficient to prove that $\lim_{n\to\infty} h_n B_{k,n} = 0$.

$$\begin{aligned} h_n^2 B_{k_1 n}^2 &\leqslant (1+\epsilon'_{k_1})^2 k_1^2 h_n^2 n^{-2a} \\ &\leqslant \frac{(1+\epsilon'_{k_1})^2}{a} k_1^{2a} n^{-2a} \left(\sum_{k=k_1}^{n} ak^{-1} B_{k,n} c_k^{-2} k^{a-1} n^{-a} \right)^{-1} \\ &= \text{constant} \left(\sum_{k=k_1}^{n} ak^{-1} B_{k,n} c_k^{-2} k^{a-1} n^{a} \right)^{-1}. \end{aligned} \tag{1}$$

If $a > \frac{1}{2}$ then $n^a k^{a-1} c_k^{-2} > c^{-2} k^{2a-1}$ which goes to ∞ as $k \to \infty$ and hence by lemma 11, the last term in (1) goes to 0 as $n \to \infty$.

LEMMA 13 (Sacks). *Let $a > \frac{1}{2}$ and suppose that $c_k \leqslant c < \infty$ for all k. Let $\{\delta_k\}$ be a sequence of real numbers converging to δ where δ may be ∞. Then if k_0 is a fixed positive integer*

$$\lim_{n\to\infty} h_n^2 \sum_{k=k_0}^{n} a^2 c_k^{-2} k^{-2} B_{k,n}^2 \delta_k = \delta.$$

Proof. The proof is easily accomplished upon noting that for any fixed k_1

$$\lim_{n\to\infty} h_n^2 \sum_{k=k_1}^{n} a^2 k^{-2} c_k^{-2} B_{kn}^2 = 1.$$

LEMMA 14 (Sacks). *Let $\{d_k\}$ be a sequence of positive numbers such that*

$$\frac{d_k}{d_{k+1}} = 1 + \epsilon_{k/k}, \quad \textit{where } \epsilon_k \to 0 \textit{ as } \quad k \to \infty. \tag{1}$$

Let $q > -1$. Then for any positive integer k_0,

$$\sum_{k=k_0}^{n} d_k k^q \approx (1+q)^{-1} d_n n^{q+1}.$$

Proof. Since $\sum_{k=1}^{n} k^q \, (1+q)^{-1} n^{p+1}$ what we have to show will be accomplished if we show that

$$\frac{\sum_{k=1}^{n} d_k k^q}{d_n \sum_1^n k^q} - 1 = \frac{\sum_{k=1}^{n-1} (d_k - d_{k+1}) \sum_{j=1}^{k} j^q + d_n \sum_{j=1}^{n} j^q}{d_n \sum_1^n k^q} - 1$$

$$= \frac{\sum_{k=1}^{n-1} \left(\frac{d_k}{d_n} - \frac{d_{k+1}}{d_n}\right)\left(\sum_{j=1}^{k} j^q\right)}{\sum_1^n k^q}$$

$$= \frac{\sum_{k=1}^{n-1} \frac{d_{k+1}}{d_n}\left(\frac{d_k}{d_{k+1}} - 1\right)\left(\sum_{j=1}^{k} j^q\right)}{\sum_1^n k^q} = A_n,$$

$$A_n \to 0 \quad \text{as} \quad n \to \infty.$$

Using (1) we see that for n sufficiently large

$$d_n = d_1 \prod_{j=1}^{n-1} d_{j+1} d_j^{-1} \geqslant \text{constant} \exp\left[\sum_{j=1}^{n-1} \epsilon_{j/j}\right]$$

$$\geqslant \text{constant} \exp\left[-\epsilon \sum_{j=k_1}^{n-1} \frac{1}{j}\right] \geqslant \text{constant}\, k_1^{\epsilon} n^{-\epsilon},$$

where k_1 is chosen so that $|\epsilon_j| < \epsilon < q+1$ for all $j \geqslant k_1$. Thus $d_n n^{q+1} \to \infty$ as $n \to \infty$, and, therefore, in order to show that $A_n \to 0$, we can start the outer sum in the numerator of A_n at $k = k_1$.

By use of (1) we have, for $n > k \geqslant k_1$,

$$d_{k+1} d_n^{-1} = \prod_{j=k+1}^{n-1} d_j d_{j+1}^{-1} \leqslant \prod_{j=k+1}^{n-1} (1 + \epsilon j^{-1}) \leqslant \text{constant } n^{\epsilon} k^{-\epsilon}$$

and

$$|d_k d_{k+1}^{-1} - 1| \leqslant \epsilon k^{-1}.$$

Also A_n must tend to 0 since for all n

$$\frac{\sum_{k=k_1}^{n} \text{constant } n^{\epsilon} k^{-\epsilon} \epsilon k^{-1} \sum_{j=1}^{k} j^q}{\sum_{k=1}^{n} k^q}$$

$$\leqslant \epsilon \text{ constant } n^{\epsilon - q - 1} \sum_{k_1}^{n} k^{-\epsilon - 1} k^{q+1} \leqslant \epsilon \text{ constant}$$

and ϵ is arbitrary. Note that if $\{d_k\}$ satisfies (1) then $\{d_k^\rho\}$ also satisfies (1) for any real number ρ.

LEMMA 15 (Sacks). *Let* $c_k = d_k k^{-r}$, *where* $r \geqslant 0$ *and where* $\{d_k\}$ *satisfies* (1) *of lemma* 14. *Let* a *be a real number greater than* $\frac{1}{2}$. *Then for any positive* k_0 *and any positive number* ρ

$$\sum_{k=k_0}^{n} a^2 c_k^{-\rho} k^{-2} B_{k,n}^2 \approx a^2 (2a+r\rho-1)^{-1} c_n^{-\rho} n^{-1} \quad as \quad n \to \infty. \quad (1)$$

In particular, if $k_0 = 1$ *and* $\rho = 2$ (1) *becomes*

$$h_n^2 \approx a^{-2}(2a+2r-1)\, nc_n^2. \quad (2)$$

Proof. Let $\epsilon > 0$ and let k_1 be large enough so that in problem (1) $\epsilon_k' < \epsilon$ for $k > k_1$. Then using problem (1) and lemma 14 the conditions of lemma 14 are satisfied if one takes into account the conditions stated here and the remarks following the statement of lemma 14, we obtain

$$\begin{aligned} a^2 \sum_{k=k_0}^{n} c_k^{-\rho} k^{-2} B_{kn}^2 &\leqslant (1+\epsilon)\, a^2 \sum_{k_1}^{n} c_k^{-\rho} k^{2a-2} n^{-2a} + (n^{-2a}) \\ &= (1+\epsilon)\, a^2 \sum_{k_1}^{n} d_k^{-\rho} k^{2a+\rho r-2} n^{-2a} + O(n^{-2a}) \\ &\leqslant (1+\alpha_n)\, a^2 (2a+\rho r-1)^{-1} d_n^{-\rho} n^{2a+\rho r-1} n^{-2a} \\ &\quad + O(n^{-2a}) \\ &= (1+\alpha_n)\, a^2 (2a+\rho r-1)^{-1} c_n^{-\rho} n^{-1} + O(n^{-2a}), \end{aligned}$$

where $\alpha_n \to \epsilon$ as $n \to \infty$. Similar calculations produce

$$a^2 \sum_{k_0}^{n} c_k^{-p} k^{-2} B_{k,n}^2 \geqslant (1-\alpha_n)\, a^2 (2a+\rho r-1)^{-1} c_n^{-p} n^{-1}.$$

Since $n^{1-2a} c_n^p \to 0$ as $n \to \infty$ and since ϵ is arbitrary we have obtained

$$\sum_{k=k_0}^{n} a^2 c_k^{-p} k^{-2} B_{k,n}^2 \approx a^2 (2a+r\rho-1)^{-1} c_n^{-p} n^{-1}$$

for sufficiently large n.

3. Other inequalities

Lemma 16. *If* $0 < a < b$ *and* $0 < t < 1$ *then*

$$a^{1-t}b^t \leqslant (1-t)\,a+tb.$$

Proof. If $a = b$ then both sides are equal. We want to show that

$$\frac{a^{1-t}b^t}{(1-t)\,a+tb} < 1,$$

i.e.
$$\left[\frac{b}{(1-t)\,a+tb}\right]^t < \left[\frac{(1-t)\,a+tb}{a}\right]^{1-t}$$

or
$$t\log\frac{b}{(1+t)\,a+tb} < (1-t)\log\frac{(1-t)\,a+tb}{a}.$$

Left-hand side $= t$

$$\int_{(1-t)\,a+tb}^{b}\frac{1}{x}dx < t\left[\frac{b-(1-t)\,a-tb}{(1-t)\,a+tb}\right] = \frac{t(1-t)\,(b-a)}{(1-t)\,a+tb}.$$

Right-hand side $= (1-t)$

$$\int_{a}^{(1-t)\,a+tb}\frac{1}{x}dx > (1-t)\left[\frac{(1-t)\,a+tb-a}{(1-t)\,a+tb}\right] = \frac{t(1-t)\,(b-a)}{(1-t)\,a+tb}$$

$$> \text{left-hand side.}$$

Lemma 17. *If* X *is a random variable* $E[|X|^r]^{1/r}$ *is a non-decreasing function of* r, *for* $r > 0$.

Proof. Suppose $0 < p < q$, we want to show that

$$[E|X|^p]^{1/p} \leqslant [E|X|^q]^{1/q}.$$

Consider

$$\frac{[E|X|^p]^{1/p}}{[E|X|^q]^{1/q}} = \left[E\left\{\frac{|X|^p}{[E|X|^q]^{p/q}}\right\}\right]^{1/p}$$

$$= \left[E\left\{\left(\frac{|X|^q}{[E|X|^q]}\right)^{p/q}\right\}\right]^{1/p}$$

$$\leqslant \left[E\left\{\frac{p}{q}\frac{|X|^q}{E|X|^q}+\frac{q-p}{q}\right\}\right]^{1/p} = (1)^{1/p} = 1.$$

This follows from lemma 16 and the fact that if $X \leqslant Y$ then $E(X) \leqslant E(Y)$. Hence

$$[E|X|^p]^{1/p} \leqslant [E|X|^q]^{1/q}.$$

LEMMA 18. *If* $0 < t < 1$ *then*

$$|EXY| \leqslant [E|X|^{1/1-t}]^{1-t}[E|Y|^{1/t}]^t$$

provided non-negative expectations exist.

Proof. If the right-hand side (r.h.s.) is zero, result is trivial. Suppose r.h.s. $\neq 0$ then above inequality is true if

$$E[|X||Y|] \leqslant \text{r.h.s.} \Leftrightarrow E\left\{\left(\frac{|X|^{1/1-t}}{E|X|^{1/1-t}}\right)^{1-t}\left(\frac{|Y|^{1/t}}{E|Y|^{1/t}}\right)^t\right\} \leqslant 1.$$

But $$\text{r.h.s.} \leqslant E(1-t)\left\{\frac{|X|^{1/1-t}}{E|X|^{1/1-t}}+t\frac{|Y|^{1/t}}{E|Y|^{1/t}}\right\} = 1.$$

Hence the desired result.

LEMMA 19. *If each of a and b is a real number and $r > 1$ then*

$$[|a|+|b|]^r \leqslant 2^{r-1}[|a|^r+|b|^r].$$

Proof. Let x be a random variable such that

$$P[X = |a|] = \tfrac{1}{2},$$

$$P[X = |b| = \tfrac{1}{2},$$

then $$E|X| = \frac{|a|+|b|}{2},$$

$$E|X|^r = \frac{|a|^r+|b|^r}{2},$$

$$\frac{|a|+|b|}{2} = E|X| \leqslant [E|X|^r]^{1/r} = \left[\frac{|a|^r+|b|^r}{2}\right]^{1/r},$$

$$\left[\frac{|a|+|b|}{2}\right]^r \leqslant \frac{|a|^r+|b|^r}{2},$$

$$[|a|+|b|]^r \leqslant 2^{r-1}[|a|^r+|b|^r].$$

4. Problems

1. Let $\{a_n\}$ be a sequence of positive real numbers such that

$$\sum_n a_n = \infty, \quad \sum_n a_n^2 < \infty$$

and

$$B_{k,n} = \prod_{j=k+1}^{n} (1-a_j) \qquad 0 \leqslant k < n,$$
$$= 1 \qquad k = n,$$

then prove that

$$(1-\epsilon_k)\exp\left(-\sum_{j=k+1}^{n} a_j\right) \leqslant B_{k,n} \leqslant (1+\epsilon_k)\exp\left(-\sum_{j=k+1}^{n} a_j\right)$$

for all $n \geqslant k$, where $\epsilon_k \to 0$ as $k \to \infty$. In particular, if for $a > 0$, $a_n = an^{-1}$ then

$$(1-\epsilon_k')\, k^a n^{-a} \leqslant B_{k,n} \leqslant (1+\epsilon_k')\, k^a n^{-a},$$

where $\epsilon_k' \to 0$ as $k \to \infty$. (Sacks)

2. Let $\{b_n\}$, $\{c_n\}$, $\{d_n\}$, be real number sequences such that

$$b_n > 0 \quad \text{and} \quad \sum_n b_n < \infty, \tag{1}$$

$$c_n \geqslant 0 \quad \text{and} \quad \sum_n c_n = \infty, \tag{2}$$

$$\Sigma\, d_n = \infty. \tag{3}$$

Then

(*a*) if $\{\xi_n\}$ is a sequence of non-negative numbers such that for some integer n_0 and all $n > n_0$

$$\xi_{n+1} \leqslant \max(a, (1+b_n)\xi_n + d_n - c_n), \tag{4a}$$

where $a > 0$ then

$$\limsup_{n\to\infty} \xi_n \leqslant a; \tag{5}$$

(*b*) if, instead of (4*a*), $\{\xi_n\}$ satisfies

$$\xi_{n+1} \leqslant \max(a, (1+b_n)\,\xi_n + d_n), \tag{4b}$$

then we can conclude that the sequence $\{\xi_n\}$ is bounded.

Hint: Prove that (3) and (1) imply $\sum_{j=k}^{n} B_{j,n} d_j$ converges as $n \to \infty$ for any fixed k, where bB_{jn} is defined in lemmas and proof is exactly the same as of lemma 5. (Venter)

3. Let $\{\xi_n\}$ be a sequence of positive real numbers such that for all n large ènough,

$$\xi_{n+1} \leqslant \left(1-\frac{c_n}{n}\right)\xi_n + \frac{d}{n^{1+p}},$$

where $d > 0$ and $c_n \to c$ as $n \to \infty$. Then prove

if $\qquad c > p > 0, \quad \xi_n = O(n^{-p}),$

if $\qquad c = p > 0, \quad \xi_n = O(n^{-c \log n}),$

if $\qquad p > c > 0, \quad \xi_n = O(n^{-c}).$ (Venter)

4. Suppose $A > 0$, $\{h_n\}$ $\{b_n\}$ are sequences of real numbers such that

$$b_{n+1} = (1-An^{-1})\,b_n + n^{-1}h_n.$$

Then prove that

$$b_n \to 0 \quad \text{if and only if} \quad \frac{1}{n}\sum_{j=1}^{n} h_j \to 0.$$

Hint:

$$\sum_{j=1}^{n} h_j = (A-1)\sum_{j=1}^{n} b_j + nb_{n+1}.$$

(Fabian)

5. Let $b_n, A_n, D_n, \alpha, \beta, B$ be real numbers, let

$$a_0 = \liminf A_n \quad \text{and} \quad a_1 = \limsup A_n$$

be finite and

$$b_{n+1} = b_n(1-A_n n^{-\alpha}) + Bn^{-\alpha-\beta} + D_n \tag{1}$$

for all sufficiently large n and $0 < \alpha \leqslant 1$, $0 < \beta$, $0 < B$. Let $C_i = a_i$ if $\alpha < 1$ and $C_i = a_i - \beta$ all $\alpha = 1$.

If the D_n's are non-positive and $C_0 > 0$ then prove that

$$\limsup n^\beta b_n \leqslant B/C_0; \tag{2}$$

if they are non-negative and $C_1 > 0$ then prove that

$$\liminf n^\beta b_n \geqslant B/C_1. \quad \text{(Fabian)} \tag{3}$$

6. Let b_n be numbers satisfying

$$b_{n+1} \leqslant b_n(1-A_n^{-1}) + B_n n^{-\beta},$$

where $A > \beta$, $\beta > 0$, and $\sum_{n+1}^{\infty} B_n$ converges. Then prove that $\limsup n^\beta b_n < +\infty$. (Fabian)

BIBLIOGRAPHY

[1] Becnenbach, E. F. and Bellman, R. *Inequalities.* Springer-Verlag, Berlin, 1961.

[2] Bertram, J. E. *Control by Stochastic Adjustment.* AIEE Applications and Industry Report (1960).

[3] Block, H. D. Estimates of error for two modifications of the Robbins–Monro stochastic approximation methods. *Ann. Math. Stat.* **28** (1957), 1003–1010.

[4] Blum, J. R. Approximation methods which converge with probability one. *Ann. Math. Stat.* **25** (1954), 382–386.

[5] Blum, J. R. Multidimensional stochastic approximation method. *Ann. Math. Stat.* **25** (1954), 737–744.

[6] Blum, J. R. A note on stochastic approximation. *Proc. Amer. Math. Soc.* **9** (1958), 404–407.

[7] Box, G. E. P. and Jenkins, G. M. Some statistical aspects of adoptive optimization and control. *J. R. Stat. Soc.* Series B, **24** (1962), 297–343.

[8] Box, G. E. P. and Wilson, K. B. On the experimental attainment of optimum conditions. *J. R. Stat. Soc.* Series B, **13** (1951), 1–45.

[9] Brownlee, K. A., Hodges, J. L. and Rosenblatt, M. The up-and-down method with small samples. *Journal of Amer. Stat. Assoc.* **48** (1953), 262–277.

[10] Burkholder, D. L. On class of stochastic approximation procedures. *Ann. Math. Stat.* **27** (1956), 1044–1059.

[11] Cameron, R. H. and Martin, W. T. The orthogonal development of nonlinear functionals in series of Fourier–Hermite functionals. *Ann. Math.* **48** (1947), 385–392.

[12] Chung, K. L. On stochastic approximation methods. *Ann. Math. Stat.* **25** (1954), 463–483.

[13] Cochran, W. G. and Davis, M. *Sequential experiments for estimating the median lethal dose.* Colloques internationaux du centre National de la Recherche Scientifique, no. 110 (1963), 181–194.

[14] Cochran, W. G. and Davis, M. Stochastic approximation to the median effective dose in bioassay. *In Stochastic Models in Medicine and Biology*, edited by John Gurland, Madison: University Wisconsin Press, (1964).

[15] Comer, J. P. Jr. Some stochastic approximation procedures for use in process control. *Ann. Math. Stat.* **35** (1964), 1136–1146.

[16] Comer, J. P. Jr. Application of stochastic approximation to process control. *J. Roy. Stat. Soc.* Series B, **27** (1965), 321–331.

[17] DAVENPORT, W. B. Jr. and ROOT, W. L. *An Introduction to the Theory of Random Signals and Noise.* McGraw-Hill Book Co., New York (1958).

[18] DAVIS, M. *Sequential Experiments in Bioassay.* Ph.D. Thesis, Harvard University (1965).

[19] DERMAN, C. An application of Chung's Lemma to the Kiefer-Wolfowitz stochastic approximation procedure. *Ann. Math. Stat.* **27** (1956), 532–536.

[20] DERMAN, C. Non-parametic up-and-down experimentation. *Ann. Math. Stat.* **28** (1957), 795–797.

[21] DERMAN, C. and SACKS, J. On Dvoretzky's stochastic approximation theorem. *Ann. Math. Stat.* **30** (1959) 601–606.

[22] DOOB, J. L. *Stochastic Processes.* John Wiley, New York, (1953).

[23] DIXEN, W. J. and MOOD, A. M. A method for obtaining and analyzing sensitivity data. *Journal of Amer. Stat. Assoc.* **43** (1948), 109–126.

[24] DRIML, M. and HANS, O. *Continuous stochastic approximations.* Transactions of the second Prague Conference on Information Theory, Statistical Decision Functions and Random Processes (Czechoslovak Academy of Sciences). Prague 1960, 113–122.

[25] DRIML, M. and NEDOMA, J. *Stochastic approximations for continuous random processes.* Transactions of the second Prague Conference on Information Theory, Statistical Decision Functions and Random Processes (Czechoslovak Academy of Sciences). Prague 1960, 145–158.

[26] DUBINS, L. E. and FREEDMAN, D. A. A sharper form of Borel–Cantell's lemma and the strong law. *Ann. Math. Stat.* **23** (1965), 800–807.

[27] DUPAČ, V. On the Keifer–Wolfowitz approximation method. *Casopis Pest. Mat.* **82** (1957), 47–75.

[28] DUPAČ, V. A dynamic stochastic approximation method. *Ann. Math. Stat.* **36** (1965), 1695–1702.

[29] DVORETZKY, A. On stochastic approximation. *Proceedings of the third Berkeley symposium on Mathematical Statistics and Probability,* I. (1956), 39–55.

[30] EPLING, M. L. *A multivariate stochastic approximation procedure.* Technical Report No. 5, Department of Statistics, Stanford University (1964).

[31] FABIAN, V. Stochastic approximation methods. *Czechoslovak Mathematical Journal,* **10** (85), (1960), 123–159.

[32] FABIAN, V. A new one dimensional stochastic approximation method for finding a local minimum of a function. *Trans. third Prague Conf.* (1964), 85–105.

[33] FABIAN, V. On asymptotic normality in stochastic approximation. *Michigan State University Statistical Laboratory Publications* (1967).

[34] FABIAN, V. Stochastic approximation of minima with improved asymptotic speed. *Ann. Math. Stat.* **38** (1967), 191–200.

[35] FARRELL, R. H. Bounded length confidence interval for the zero of a regression function. *Ann. Math. Stat.* **33**, (1962), 237–247.

[36] FELLER, W. *An Introduction to Probability Theory and its Applications*, **1**, John Wiley, New York, 1957.

[37] FELLER, W. *An Introduction to Probability Theory and its Applications*, **2**, John Wiley, NewYork, 1966.

[38] FRIEDMAN, M. and SAVAGE, L. J. Experimental determination of the maximum of a function. *Selected Techniques of Statistical Analysis.* McGraw-Hill, New York, 1947, 363–372.

[39] FRIEDMAN, S. On stochastic approximations. *Ann. Math. Stat.* **34** (1963), 343–346.

[40] GADDUM, J. H. *Reports on Biological Standards*, III. *Methods of biological assay depending on a quantal response.* Special Report No. 183 Medical Research Council. H. M. Stationery Office, London, 1933.

[41] GARDNER, L. A. Jr. *Stochastic approximation and its application of prediction and control synthesis.* International Symposium on Nonlinear Differential Equations and Nonlinear mechanics. Academic Press, New York, 1963, 241–258.

[42] GRAY, K. B. *The application of stochastic approximation to the optimization of random circuits.* American Mathematical Society, Proceeding of Symposium in Applied Mathematics, **16** (1962), 178–192.

[43] GUTTMAN, L and GUTTMAN, R. In Illustration of the use of stochastic approximation. *Biometrics*, **15** (1959), 551–559.

[44] HANS, A. and SPACEK, A. *Random fixed point approximation by differentiable trajectories.* Tran. Second Prague Conference on Information Theory, Statistical Decisions and Random Processes (1960), 203–213.

[45] HODGES, J. L. and LEHMANN, E. L. *Two approximations to the Robbins–Monro process.* Proceedings of the third Berkeley Symposium on Mathematical Statistics and Probability, (1956), 95–104.

[46] HOTELLING, H. Experimental determination of the maximum of a function. *Ann. Math. Stat.* **12** (1941), 20–46.

[47] HUNTER, L. *Optimum checking procedures. Statistical Theory of Reliability*, edited by M. Zelen. University of Wisconsin Press, 1963.

[48] KALLIANPUR, G. A note on the Robbins–Monro stochastic approximation methods. *Ann. Math. Stat.* **25** (1954), 386–388.

[49] KESTEN, H. Accelerated Stochastic Approximation. *Ann. Math. Stat.* **29** (1958), 41–59.

[50] KIEFER, J. and WOLFOWITZ, J. Stochastic estimation of the maximum of a regression function. *Ann. Math. Stat.* **23** (1952), 462–466.

[51] KOLMOGOROV, A. M. *Foundation of the Theory of Probability.* Chelsea, New York, 1956.

[52] KRASULINA, T. P. *A note on some stochastic approximation processes.* Theory of Probability and Its Applications, **7** (1962), 108–113.

[53] KULLBACK, S. *Information Theory and Statistics.* John Wiley, New York, 1959.

[54] KUSHNER, H. J. *Adaptive techniques for the optimization of binary detection systems.* 22G-2 Lincoln Laboratory MIT, 9 November 1962.

[55] KUSHNER, H. J. *Methods for the adaptive optimization of binary detection systems.* 22G-2 Lincoln Laboratory MIT, 1962.

[56] KUSHNER, H. J. Hill climbing methods for the optimization of multiparameter noise disturbed systems. *Journal of Basic Engineering* (1963), 157–164.

[57] KUSHNER, H. J. A note on the maximum sample excursions of stochastic approximation processes. *Ann. Math. Stat.* **37**, (1966), 513–516.

[58] LOÈVE, M. *Probability Theory,* Van Nostrand, New York, 1960.

[59] NARAYANA, T. V. *Sequential Procedures in Probit Analysis.* Unpublished Ph.D. thesis, University of North Carolina (1953).

[60] ODELL, P. L. *An Empirical study of three stochastic approximation techniques applicable to sensitivity testing.* NAVWEPS Report 7837 (1961).

[61] OSTROWSKI, A. M. *Solution of Equations of Systems of Equation.* Academic Press, New York, 1960.

[62] PERRIN, E. B. *Estimation of parameters in systems related to the observation by an unknown monotone transformation.* Technical Report No. 1, Stanford University (1960).

[63] ROBBINS, H. and MONRO, S. A stochastic approximation method. *Ann. Math. Stat.* **22** (1951), 400–407.

[64] SACKRISON, D. J. *Application of stochastic approximation methods to system optimization.* Technical Report 391 Research Laboratory of Electronics MIT (1962).

[65] SACKRISON, D. J. *Iterative design of optimum filters for non-mean square error criteria.* Trans. of I.E.E.E. Professional Group on Information Theory, No. 3 (1963), 161–167.

[66] SACKRISON, D. J. A continuous Kiefer–Wolfowitz procedure for random processes. *Ann. Math. Stat.* **35** (1964), 590–599.

[67] SACKS, J. Asymptotic Distribution of Stochastic Approximation Procedures. *Ann. Math. Stat.* **29** (1958), 373–405.

[68] SCHALKWIJK, J. P. M. *Coding schemes of additive noise channels with feedback.* Scientific Report No. 10, Stanford Electronics Laboratories, Stanford University (1965).

[69] SCHMETTERER, L. *Stochastic approximation.* Proceedings fourth Berkeley Symposium on Mathematical Statistics and Probability, **1** (1960), 587–609.

[70] TRAUB, J. F. *Iterative Methods for the Solution of Equation.* Prentice-Hall Inc., Englewood, N.J. 1964.

[71] Venter, J. H. and Gastwirth, J. L. *Adaptive statistical procedures in reliability and maintenance problems.* Technical Report No. 99, Department of Statistics, Stanford University (1964).

[72] Venter, J. H. *On Stochastic Approximation Methods.* Ph.D. Thesis University of Chicago (1963).

[73] Venter, J. H. On Dvoretzky stochastic approximation Theorems. *Ann. Math. Stat.* **37** (1966), 1534–1544.

[74] Venter, J. H. An extension of the Robbins–Monro procedure. *Ann. Math. Stat.* **38** (1967), 181–190.

[75] Venter, J. H. On convergence of the Kiefer–Wolfowitz approximation procedures. *Ann. Math. Stat.* **38** (1967), 1031–1036.

[76] Wasan, M. T. *A technique of stochastic approximation.* Proc. Functional Analysis Symposium, QPPAM-10 (1967), 242–247.

[77] Wasan, M. T. On stochastic approximation methods. *Abstract Ann. Math. Stat.* **36** (1965), 1078.

[78] Wasan, M. T. *Stochastic approximation and its application.* American Mathematical Society Summer Seminar on Mathematics of decision sciences (1967).

[79] Wasan, M. T. Sequential optimum procedures for unbiased estimation of a binomial parameter. *Technometrics*, **6** (1964), 259–272.

[80] Wasan, M. T. *Parametric Estimation.* (Textbook.) To be published by McGraw-Hill Book Co., New York (approximate publication date Fall 1969).

[81] Wetherill, G. B. Sequential estimation of quantal response curves. *J. Roy. Stat. Soc.* **25**, Series B (1963), 1–48.

[82] Wetherill, G. B. *Sequential Methods in Statistics.* Methuen, London, 1966.

[83] Wolfowitz, J. On stochastic approximation methods. *Ann. Math. Stat.* **27** (1956), 1151–1156.

[84] Wolfowitz, J. On the stochastic approximation method of Robbins and Monro. *Ann. Math. Stat.* **23** (1952), 457–461.

INDEX

Printed in the United States
By Bookmasters